Food and Power in Hawai‘i

FOOD IN ASIA AND THE PACIFIC

Series Editors:

Christine R. Yano and Robert Ji-Song Ku

Food and Power in Hawai'i

VISIONS OF FOOD DEMOCRACY

Edited by Aya Hirata Kimura and Krisnawati Suryanata

University of Hawai'i Press | Honolulu

Printed in the United States of America

21 20 19 18 17 16 6 5 4 3 2 1

Library of Congress Cataloging-in-Publication Data

Names: Kimura, Aya Hirata, editor. | Suryanata, Krisnawati, editor.
Title: Food and power in Hawai'i : visions of food democracy / edited by Aya Hirata Kimura and Krisnawati Suryanata.
Other titles: Food in Asia and the Pacific.
Description: Honolulu : University of Hawai'i Press, [2016] | Series: Food in Asia and the Pacific | Includes bibliographical references and index.
Identifiers: LCCN 2016011179 | ISBN 9780824858537 (cloth ; alk. paper)
Subjects: LCSH: Food industry and trade—Hawaii. | Food security—Hawaii. | Agriculture—Hawaii.
Classification: LCC HD9007.H3 F66 2016 | DDC 338.1/9969—dc23 LC record available at https://lccn.loc.gov/2016011179

University of Hawai'i Press books are printed on acid-free paper and meet the guidelines for permanence and durability of the Council on Library Resources.

Contents

Acknowledgments *vii*

Introduction *1*
AYA HIRATA KIMURA AND KRISNAWATI SURYANATA

1 Tangled Roots: The Paradox of Important Agricultural Lands in Hawai'i *17*
KRISNAWATI SURYANATA AND KEM LOWRY

2 Food Security in Hawai'i *36*
GEORGE KENT

3 Kaulana O'ahu me he 'Āina Momona *54*
LILIKALĀ K. KAME'ELEIHIWA

HI'ILEI KAWELO

4 Farmers' Markets in Hawai'i: A Local/Global Nexus *85*
MONIQUE MIRONESCO

5 Is the Transgene a Grave? On the Place of Transgenic Papaya in Food Democracy in Hawai'i *116*
NEAL K. ADOLPH AKATSUKA

6 Seeds of Contestation: The Emergence of Hawai'i's Seed Corn Industry *138*
BENJAMIN SCHRAGER AND KRISNAWATI SURYANATA

7 Farming on the Margin: Women Organic Farmers in Hawai'i *156*
AYA HIRATA KIMURA

Michelle Galimba

8 Labor of Meaning, Labor of Need: Organic Farm Volunteering in Hawai'i *185*
MARY MOSTAFANEZHAD, KRISNAWATI SURYANATA, SALEH AZIZI, AND NICOLE MILNE

Chris Robb

Dean Okimoto

9 Epilogue *211*
AYA HIRATA KIMURA AND KRISNAWATI SURYANATA

Contributors 215
Index 219

Acknowledgments

We would like to thank numerous people and organizations that were critical in making this book a reality. We are grateful to the authors who agreed to provide chapters for this volume, and then waited with considerable patience, as it went through several iterations. Our thanks go to Michelle Galimba, Hiʻilei Kawelo, Dean Okimoto, and Chris Robb who shared their visions on our current food system in the narratives that enrich this volume. We would also like to thank numerous people who have engaged us in productive conversations throughout the development of this project. Finally, we appreciate feedback from two anonymous reviewers and the steady support of Masako Ikeda and Nadine Little at the University of Hawaiʻi Press who shepherded this project.

Food and Power in Hawaiʻi

Introduction

Aya Hirata Kimura and
Krisnawati Suryanata

Food is an intimate part of our daily lives. It is a basic consumption item that provides calories, nutrition, and satisfaction; is central to our social relationships; and embeds cultural, economic, and environmental elements of a place. Consequently, it is a focal point for public policy debate in many parts of the world. In Hawai'i, contentious issues such as food insecurity, food localization, and the use of genetically modified organisms (GMOs) have been in the forefront of social activism and cultural mobilization. Yet little has been written about Hawai'i's food politics and the diverse ways Hawai'i's citizens are engaged in shaping their food system. This volume presents case studies from the state to critically examine several initiatives in the agriculture and food sector. They range from legislative efforts to protect agricultural lands, to innovative ways to help small farmers, to social activism to oppose GMOs. While many initiatives are framed in the rhetoric of food localization, we view them as valuable not simply because they contribute to the local food supply or provide alternative market arrangements, but because they engage and empower the citizens to participate in food democracy. In linking the debate on food explicitly to the issue of power and democracy, we seek to reframe the discourse that has focused on increasing the amount of locally grown food or protecting local farms into the broader objectives of social justice, ecological sustainability, and economic viability.

We situate the urge to increase local food production in the broader concern that arises from the growing alienation that farmers and consumers experience in food production. In spite of the abundant food efficiently delivered by the global agro-food industry, our collective sense of vulnerability to losing our food security has increased, especially when problems such as climate change, biofuel boom, and commodity speculation led to higher food prices in 2008 (e.g., Headey and Fan 2008; Lobell et al. 2008; van der Ploeg 2010). In Hawai'i, concern about food security was amplified by reports published by researchers at the Rocky Mountain Institute, the University of Hawai'i, and the state Department of Agriculture (Leung and Loke 2008; Page, Bony, and Schewel 2007) that highlighted Hawai'i's high dependency on imported food. This heightened worry converged with a latent concern over the continuing slide of the state's agricultural industry and the rapid transformation of land use occurring on agricultural lands.

We join a growing number of analysts who caution against uncritical celebration of localism (e.g., DuPuis and Goodman 2005; Born and Purcell 2006; Morgan 2010; deLind 2011). Local is often used as a surrogate for other socially constructed and value-laden concepts including healthy, fresh, authentic, and traditional that denote quality. But as Ilbery and Kneafsey (2000) argue, quality should not merely concern the product and its composition; it should also concern the sociocultural context within which the product is bought and sold. Indeed, being *local* is not inherently desirable or *just*. The contradiction was made evident in Hawai'i when proprietors of a large farm that produces fresh vegetables for local consumption were suspected of forcing immigrant workers from Thailand to farm under inhumane conditions. Some observers, including former governors and community leaders, ironically made an argument for leniency of the sentence on the ground of their contribution to Hawai'i's food self-sufficiency (Downes 2010).

Historically, Hawai'i's agrarian economy has pursued an export-oriented strategy for at least a century. For a majority of the population food security has been largely provided through the working of the markets. Nobel laureate Amartya Sen (1981, 1995) has persuasively argued that public actions to improve food security are not limited to increasing the food supply. Equally if not more important are those actions that enhance the entitlement structures of the general populace—by securing economic opportunities (within and without agriculture) coupled with meaningful political participation—that allow people to access quality food.

It is for this reason that this book is subtitled "Visions of Food Democracy." We seek to move our discussion beyond the narrow preoccupation with local food production and its economic value, towards examining the variety of ways that citizens of Hawai'i become engaged in food issues. Drawing on work on food democracy (Hassanein 2003; Lang and Heasman

2004) and a related concept of "food citizenship" (Welsh and MacRae 1998; deLind 2002; Wilkins 2005), we urge readers to think about food issues not simply as a matter of food production per se, but as a matter of politics that are fundamentally about social, cultural, and economic powers. Neva Hassanein (2003) captures the argument for food democracy in saying: "at the core of food democracy is the idea that people can and should be actively participating in shaping the food system" and "food democracy is about citizens having the power to determine agro-food policies and practices locally, regionally, nationally, and globally" (79).

A Short History of Hawai'i's Agro-food Systems

When Captain Cook came ashore in 1778, the Hawaiian society already had a well-developed knowledge system to manage the islands' natural resources for supporting its people. As in most subsistence societies of the eighteenth century, outside food was not available. The Ahupua'a system—land divisions that run from the forested mountains to the waters off shore—was a basic building block of food self-sufficiency (Kame'eleihiwa, this volume). Following Western contact, Hawai'i's economy was gradually incorporated into the world market. From the 1830s Hawai'i began to export commodities such as sugar, molasses, bananas, coffee, oranges, potatoes, yams, salted beef, and cattle hides. The California gold rush brought a brief agricultural boom to Hawai'i in the early 1850s before it lost the market to mainland growers. Rice cultivation grew when the number of workers from China rapidly increased in Hawai'i and California, reaching its peak in the late 1880s when more than 9,000 acres were planted with rice (Philipp 1953).

In 1848, the Hawaiian royalty adopted a series of policy changes called the Māhele, which led to a good portion of the land passing into the hands of white colonists by lease, sale, or marriage (Kame'eleihiwa 1992, this volume). This acquisition of large acreages of land, coupled with the expanding market on the West Coast of the United States, facilitated the rise of sugarcane plantations on the islands in the second half of the nineteenth century. When the United States annexed the islands in 1898, the Hawaiian sugar industry was assured of access to a large protected market, and growing sugarcane quickly became the main agricultural activity in Hawai'i—including small independent growers who grew sugarcane under contract with the plantations (Philipp 1953). Within a few decades, pineapple growing and canning became Hawai'i's second major industry, reaching its peak in the mid-1950s. At this time, sugarcane and pineapple plantations occupied more than 300,000 acres of land in the state and directly and indirectly employed about one-third of the civilian workforce. These two crops accounted for more than 90 percent of the field crops (Hitch 1992).

With the increased specialization and export-oriented strategy in agriculture, the amount of food grown for local consumption stagnated or declined. The variety of fruits and vegetables were primarily grown in backyards for home use, and the dependency on imported food gradually increased. By the end of the nineteenth century, Hawai'i was already a net importer of oranges, potatoes, and cabbages. Rice production declined in the face of competition against low-cost mechanized rice growers of California, and was virtually eliminated by the mid-twentieth century. Growth in the shipping industry and the development of refrigeration technology further increased the variety and amount of imported food, and by 1952 only one-third of the food was locally grown (Philipp 1953).

In the 1970s, high labor costs and foreign competition began to diminish the profitability of the plantation industry in Hawai'i and a number of plantations were forced to close down. The legacy of the plantation economy and limited domestic markets led many producers in island nations, such as those in the Caribbean, into continuing the export-oriented strategy while diversifying into exotic tropical islands' products and fresh fruits and vegetables (McElroy and Albuquerque 1990; Raynolds 1997). However, new export crops from Hawai'i such as macadamia nuts and tropical flowers soon ran into the same problems of global competition that led to falling profit margins (Suryanata 2000, 2002).

When the sugarcane and pineapple plantations closed operations, many planners and food activists saw the increased availability of lands as an opportunity to reduce the islands' dependency on imported food. But the shift towards more domestic food production has turned out to be more difficult than imagined. First, vertically integrated food companies based outside Hawai'i, with their extensive networks of growers, processors, transportation agents, and retailers, deliver food products across vast distances and dominate the local market. Second, because of the isolated geographic location and small population, growers of generic food products in Hawai'i continue to face a "pocket market" problem (cf. Peters, Reed, and Creek 1954), a situation in which local producers do not really have the option of selling their products outside Hawai'i when global supplies are large enough to depress local prices (Suryanata 2002).

The preceding discussion shows that the structural disadvantage of locally produced food is rooted in Hawai'i's geography and economic history. Hawai'i's low self-sufficiency in food is neither a recent phenomenon, nor could it be remedied by simply substituting the plantation crops with food crops. It is important to understand this context while we push forward programs and policies to promote food democracy in Hawai'i.

Alternative Agro-food Networks in Hawai'i

Alternative agro-food networks focus on promoting ways of food provisioning that are not dependent on the conventional agro-food industry. Increasing their prominence in the 1990s in North America, Europe, and other industrial economies, the networks are diverse while sharing a motivation to challenge the status quo of food systems that are seen as unhealthy, insecure, unstable, and unjust (Goodman 2003). Mirroring the landscape of alternative agro-food networks elsewhere, several of these movements are active in Hawai'i.

While the reasoning might vary, virtually all these movements put emphasis on the value of locally grown food. Locally grown food[1] received a boost when twelve award-winning chefs popularized Hawai'i Regional Cuisine (HRC), a culinary movement that innovatively blends Hawai'i's diverse ethnic flavors with the cuisines of the world, and takes advantage of the freshest ingredients grown locally (Henderson 1994). Catering to food connoisseurs and tourists seeking authentic Hawaiian experience, HRC has developed menus that highlight the "place" of each ingredient, such as *Maui* onion, *Hamakua* tomatoes, *Kona* Kampachi, or *Waimānalo* greens (Costa and Besio 2011). The popularity of HRC in Hawai'i's upscale resorts started off a niche market for high-value locally grown food. This was critical in sustaining a cadre of local growers who had to compete in a market dominated by cheaper imported produce (Suryanata 2002). Many of the celebrated chefs have become champions of locally grown food.

More recent campaigns have expanded the target consumer groups beyond the patrons of upscale restaurants and resorts to include local residents. They include efforts to establish farmers' markets across the islands that feed into local food campaigns such as the "Eat Local Challenge" backed by a nonprofit organization Kanu Hawaiʻi. Others organize festivals that showcase local ingredients such as the breadfruit festival on the Big Island, the kava festival on Oʻahu, and the taro festival on Māui, to name a few. There are also nonprofit organizations supporting school garden projects and farm-to-school programs, such as the ʻĀina in Schools by Kōkua Hawaiʻi Foundation, a nonprofit organization started by a musician, Jack Johnson, and his wife, Kim Johnson.

Efforts to improve the Hawaiʻi food systems are also important elements in community building, youth education, and rejuvenation of Native Hawaiian culture. For instance, the Cultural Learning Center at Kaʻala in the Waiʻanae area and Ka Papa Loʻi o Kānewai in the Hawaiʻinuiākea School of Hawaiian Knowledge at the University of Hawaiʻi have restored old taro fields and work with youth to teach them taro cultivation and broader Native Hawaiian cultural values. MAʻO Organic Farm, whose primary objective is to provide youth education and social entrepreneurship, operates an organic farm in West Oʻahu while emphasizing Hawaiian practices and values. Often featured by the media as an example of the emerging local food community,[2] MAʻO Farm is one of the most important producers of organic food on the island, supplying grocery stores, farmers' markets, upscale restaurants, and its own community-supported agriculture (CSA) network.

A number of groups have also mobilized around concerns regarding GMOs. Hawaiʻi's awareness of GMO issues got a slower start in comparison with other parts of the world. In the 1980s, researchers at Cornell University and the University of Hawaiʻi developed genetically modified papaya to counter a viral disease that was devastating the industry in Hawaiʻi. When the GM papaya was commercially released in 1998, public criticism was initially mute, although organic farmers protested the potential of gene contamination (Akatsuka, this volume). Activism against GMO became much more visible in 2006 with the plan to patent several varieties of kalo (taro) as well as the launch of GMO kalo projects at the University of Hawaiʻi. Native Hawaiian activists demanded that the university give up its patents on three lines of (non-GMO) kalo and genetic modification of kalo on the ground that kalo is a cultural property of the Hawaiians that should neither be genetically modified nor patented.

GMO presence in Hawaiʻi's agricultural landscape dramatically changed in the 2000s when corporate consolidation and advances in molecular breeding led major seed corn companies to increase their investment in tropical nurseries, including in Hawai'i (Schrager and Suryanata, this volume). The rapid growth of seed corn operations has raised concerns and drawn opposition from inside and outside the state—leading some observers to dub Hawai'i as "ground zero for GMOs" (McAvoy 2014). Today the coalition consists of activists who, in addition to Native Hawaiian rights, are concerned with issues of economic justice, consumer rights to know, environmental health, and food security.

The movements around issues of agriculture and food in Hawai'i have provided important criticisms of the existing food system from varying perspectives. They have opened a space for discussion in multiple venues, such as in the legislative arena as well as in the market place, on the role of agriculture and food in embodying sustainability and social problems while at the same time offering something to act on in a tangible way. The chapters in this book provide a critical analysis of these efforts—highlighting both the opportunities and the pitfalls—and hope to further inform the discussion.

Visions of Food Democracy: Outline of This Volume

There are eight analytical chapters in this volume that explore diverse aspects of the Hawai'i food systems. In addition, with the hope to capture on-the-ground functions and manifestations of visions for the Hawai'i food systems, we decided to solicit narratives from people who have engaged directly with food production and activism to learn from their experiences. The narratives attest to the diversity of visions and motivations of people who are actively involved in the actual making of, and activism on food. We let the contributors decide on the format of presentation, hoping that they would choose the most appropriate space to convey the spirit of what they are doing. Therefore, the narratives are told in diverse forms—some of them are essays written by food actors themselves while some are direct excerpts from interviews.

The chapter following this introduction is written by Krisnawati Suryanata and Kem Lowry, a geographer and an urban planner, who are long-term observers of agricultural and land use changes in Hawai'i. The chapter presents an overview of the different policy tools affecting agricultural and rural lands in Hawai'i, some of which created unintended consequences in their wakes. It examines a series of laws and policies aimed at preserving prime agricultural lands that many had assumed to be *the* key to protecting the agricultural sector, and by extension, to producing food for local consumption. Their analysis of Acts 183 and 233 on "important agricultural lands" shows that protecting prime agricultural lands has become an end in itself rather than a means to facilitate the achievement of a vision for a new agricultural future for Hawai'i. It has had as little impact on the processes of rural gentrification as it has had on improving food security for the islands. Like many contemporary rural regions in a developed economy, rural Hawai'i is no longer only a site of agricultural production, but also a site of landscape consumption for people seeking an alternative lifestyle. In addition, rural areas must also contend with the threat of expanding urban areas. Many competing visions are imposed on rural Hawai'i, visions that are shaped by history, culture, and political economic interests in this state. Efforts to improve food democracy and food production for local consumption must be viewed against this backdrop. Furthermore, they must be placed in the context of globalized agro-food systems as well as global capital mobility that has played a large role in the flow of investment to the state.

In the following chapter a political scientist, George Kent, challenges the uncritical pursuit of food self-sufficiency that has been rationalized as increasing the state's preparedness against shipping disruption. Kent argues that this effort might increase food's cost, and reiterates the point that local food that is currently available is not necessarily fair as low-income consumers could be sidelined in the push for food localization. He further points out that in contrast to the enthusiasm for promoting agriculture and local food production in the state, relatively little has been done in addressing food insecurity of the poor, especially by the state government. Food democracy needs to consider food security for all—particularly the poor and the marginalized.

Kent's concern about the impact of food localization on low-income communities leads to the question: what if we give means to grow more food by and for these marginalized communities? Lilikalā K. Kame'eleihiwa's chapter provides various examples of projects by Native Hawaiian groups to grow their food in a way that is consistent with their tradition and values. A Native Hawaiian historian, Kame'eleihiwa relates these projects to the long history of the Ahupua'a system, which had been devastated by centuries of plantation agriculture and the privatization of land. Interestingly, her chapter points out an important aspect of food localization that is not emphasized in Kent's criticism of local food projects—food democracy is not only about sufficiency in terms of the amount of food grown locally, but rather, more crucially, about cultural revitalization and community building. Kame'eleihiwa's chapter is accompanied by a narrative that is based on an interview with Hi'ilei Kawelo, who oversees Paepae o He'eia, a traditional Native Hawaiian fishpond restoration project. An example of Native Hawaiian efforts that ground food in culture and history, the project educates the community and the youth about sustainably managing resources to feed the people on the islands.

The tension between *farm* security and *food* security explored in Kent's chapter is again apparent in the chapter on farmers' markets on O'ahu by a political scientist, Monique Mironesco. Farmers' markets are often seen at the forefront of food localization nationwide. Not only do farmers' markets help local farms, they also provide a space for both consumers and farmers to start questioning the hegemonic system. On O'ahu, however, some farmers' markets attract a large number of tourists from the US mainland and other countries. Mironesco raises a concern that low-income families tend to be excluded from these markets.

Neal K. Adolph Akatsuka's chapter examines the case of genetically modified (GM) papayas. Akatsuka, an anthropologist, reminds us that expression of food democracy at the local level must be situated in a broader historical and political economic context. He cautions against the tendency of essentializing GM papayas as either a savior or harbinger of destruction. While both positions raise valid underlying concerns, they both tend to overlook the more difficult challenges of managing the smallholder papaya industry in a global market. On the one hand, the benefits and costs of GM papayas are unevenly distributed. On the other hand, he sees a problem in anti-GMO activists' push for "pure" agriculture, pointing out that the line between "pure" and "impure," and "natural" and "unnatural" is always ambiguous.

While GM papayas were developed in a public institution and distributed free to papaya smallholders, the GM seed corn industry in Hawai'i is dominated by major seed corporations. The chapter by Schrager and Suryanata elucidates how the seed business quickly rose to be the state's number one agricultural sector in the 2000s after a series of techno-scientific innovations and organizational restructuring in the corn seed industry. Wider application of molecular breeding technique enables seed corporations to reduce the necessity of locating seed corn nurseries in the Corn Belt. The enhanced flexibility allows seed corporations to speed up crop improvement and take advantage of locations that offer a mix of geographical advantages. Under flexible accumulation seed corporations can swiftly shift, expand, and contract their off-season and year-round nurseries, which leads us to question whether the industry can continue to provide stable employment and economic opportunities in Hawai'i.

Agriculture in the state is far from monolithic. If GMO represents capital- and technology-intensive agriculture, sociologist Aya Hirata Kimura's chapter examines the other end of the spectrum: organic agriculture practiced by women farmers. Organic agriculture diverges from conventional agriculture, not only on farming techniques and market prices but also on sociocultural dynamics. She focuses on how women are playing an increasingly significant role in organic farming. While both of them (organic agriculture and women farmers) occupy a marginal position in the state's agriculture, they articulate the important themes for food democracy—commitment to community and sustainability rather than economic profit.

Michelle Galimba is one of the leading women in agriculture in the state, and her narrative follows Kimura's chapter. She tells the story of her family, which has been intimately linked to the dynamics of a broader Hawai'i agriculture. Her paternal grandfather was part of the wave of Filipino immigrants who were brought to meet the needs of the perennially labor-short sugarcane plantations in the early twentieth century. The beginning of her ranch was tied to the demise of local dairy and livestock operations, shifting the state to increasing dependency on food imports. Yet like the creative and unconventional strategies taken by organic women farmers in Kimura's chapter, Galimba's Kuahiwi Ranch is flourishing by expanding its direct link to consumers. Her story counters a dominant image that codes agriculture and ranching as a masculine domain, leaving women with only subsidiary roles. Women can be, and in fact many women often are, in charge of agriculture, often linking to and taking leadership roles in alternative agro-food networks.

The last chapter continues to examine the challenges faced by organic farmers. One of the strategies taken by organic farmers to lower their costs is to turn to volunteers recruited via World Wide Opportunities on Organic Farms (WWOOF). Mostafanezhad, Suryanata, Azizi, and Milne question whether such volunteerism solves the fundamental problems faced by the farms in the state, such as the high cost of production and their inability to capture price premium. Their chapter also brings up a crucial characteristic of Hawai'i as a tourism state. Like in tourist-crowded farmers' markets described by Mironesco (this volume), this chapter examines how Hawai'i's organic agriculture becomes increasingly a site of tourism. Their critical look at volunteer tourism as market-based activism echoes the criticisms of ethical food consumption by many other scholars (e.g., Bryant and Goodman 2004; Morgan 2010), making it clear that food democracy needs more than an individual change in consumptive behaviors, and a sustained engagement with the issue by local communities must be included.

This chapter is followed by two interviews by Nicole Milne, who highlights the problem of securing labor in agriculture. She interviewed farmers about the challenges they faced in securing a stable work force. The first interview is with Chris Robb who operates an organic farm on the island of Hawai'i. In contrast to a number of organic farms in the state that rely on volunteers recruited through organizations such as the WWOOF, Robb Farms employs seven full-time workers with full benefits. The second interview is with Dean Okimoto of Nalo Farms. Okimoto's farm was among the pioneers who formed collaboration with the chefs of upscale restaurants to promote locally grown food. Okimoto claims that his ability to recruit and retain twenty-four workers in his farm was due to a business strategy that focused on upscale outlets and high margin produce. The labor situation in these two farms illustrates the difficult trade-off between pursuing economic justice for farm workers on the one hand, and increasing affordability of quality food for the general populace on the other.

Towards Food Democracy

The book's collection demonstrates that visions for the future of Hawai'i's food systems are diverse. Nonetheless, we point out some critical issues to consider when thinking through the pursuit of food democracy. First, many of the concerns discussed in this book are themselves the direct and indirect outcome of neoliberal policies that encourage privatization while reducing the role of government. The presence of people who are hungry and malnourished in our midst ought to be considered in tandem with the expansion of neoliberal policies. Food insecurity is linked to, for instance, the deregulation of workforce, erosion of job security and wage rates, and reduction of government role in providing safety net for the poor that have been going on since the 1980s. And it is no accident that many of the poor depend on Food Bank programs, instead of government agencies, to meet their nutritional needs.

Ironically, neoliberalism also shapes many of the initiatives to address these concerns. Some strategies emphasize providing alternative food products and raising consumer awareness as the steps towards better environment, food security, and farm livelihood. However, as others have observed (e.g., Guthman 2004; Johnston, Biro, and MacKendrick 2009), the seeming expansion of (shopping) choice belies the actual narrowing of how we imagine food activism beyond consumption action at the cash register. As Laura deLind (2002, 218) observed, "there is a danger in equating production and consumption, responsible or otherwise, with citizenship. A good producer, a good product, a good consumer is not at all the same thing as a good citizen." Furthermore, the fact that these alternative products are often more expensive than the conventional counterparts means that the opportunities for participation by low-income communities would be limited.

Most food-related initiatives in Hawai'i are sponsored by nongovernmental organizations funded by private foundations, masking the absence and decrease in publicly funded initiatives. There are several drawbacks in this strategy. Private donations and foundation money could be unreliable, particularly when economic situation turns for the worse. Projects supported by private funding tend to lack continuity and stability. Furthermore, without the official programs of the state, the distribution of the benefits of these projects might be skewed by the availability of resources and key personnel. For example, school garden projects across Hawai'i provide promising avenues for children to learn valuable lessons in farming, nutrition, and the environment. Yet the program is limited to select schools that partner with nonprofit groups, and can only reach a limited number of beneficiaries. To reach broader constituents, these nongovernmental efforts must be coupled with policy endeavors and public programs so as to benefit all citizens.

Second is the need to reframe the debate on the future of agricultural and rural lands. Some critics of development projects have deployed the rhetoric of local food production and agricultural preservation to counter proposals for the development of housing and commercial use on agricultural lands. Yet land use conversion in Hawai'i continues to be driven by logics that are only minimally related to agriculture (Suryanata and Lowry, this volume). It is important to recognize that not all local farms are created equal. To some, farming is simply an economic enterprise motivated primarily by private accumulation. Other benefits that may accrue to the community—such as food security, community pride, or global energy saving—are welcome bonuses, but not central to these farmers. Others view farming as fulfilling a bigger mission not comparable to other ways of making a living. Most farms fall between these two extremes. Failing to recognize the diverse visions of these stakeholders leads to contradictory agro-food policies that at times pit stakeholders in volatile situations. Thus, conversations on land use would benefit from attending to a wider range of issues including urban sprawl, climate change threats, loss of wild habitat and biodiversity, as well as stress on water resources and other public infrastructure.

Third, actions to promote food democracy must attend to gender, class, and race issues in the food systems. Many Native Hawaiians still suffer from diverted water and pesticide drift that result in insufficient water flow to their lo'i (smaller land division) and damage to fish stocks from river to ocean. Food democracy in Hawai'i cannot be achieved without recognizing the historical injustices experienced by Native Hawaiian communities. In addition, with the history of agrarianism that privileges "family farms" as the cornerstone of American rural society, farm ownership is the implicit norm in agricultural policies and practices. DuPuis and Goodman (2005) observe that while the image of the family farm helped galvanize alternative agro-food networks, it often obfuscated the struggles of farm workers—many of whom are recent immigrants with their own concerns. The privileging of farm owners at the cost of farm workers is also a pertinent issue in Hawai'i. In addition, female- and minority-dominant food sectors such as food services and processing sectors have received much less attention in the discussion.

Food is a powerful medium to think about the society, the environment, and the culture. Given its rich cultural meanings, significant economic impact, and health implications, food opens up a space for interrogating many challenges and opportunities that communities face today. The chapters in this book show that each of the commonly cited objectives such as cultural preservation, food self-sufficiency, farmland protection, anti-GMO activism, or agricultural diversification must be situated in the broader context, to recognize the contradictions and potential impacts on citizen participation, social justice, and environmental sustainability. The volume as a whole adds more questions than answers to the discussion of food democracy: *how can we build a more just, culturally rooted, equitable, and democratic food system?* We hope this volume will inspire and elevate that dialogue.

Notes

1. "Locally grown food" is distinct from "local food"—a hybrid cuisine from plantation labor with ingredients that are generally not local, such as rice and canned meat (see Costa and Besio 2011 for a full discussion).

2. MA'O Organic Farm was the site of a field trip organized for First Lady Michelle Obama during the 2011 Asia Pacific Economic Cooperation (APEC) meeting in Hawai'i.

References

Born, Branden, and Mark Purcell. 2006. "Avoiding the Local Trap: Scale and Food Systems in Planning Research." *Journal of Planning Education and Research* 26, no. 2:195–207.

Bryant, Raymond L., and Michael K. Goodman. 2004. "Consuming Narratives: The Political Ecology of 'Alternative' Consumption." *Transactions of the Institute of British Geographers* 29, no. 3:344–366.

Costa, LeeRay, and Kathryn Besio. 2011. "Eating Hawai'i: Local Foods and Place-Making in Hawai'i Regional Cuisine." *Social & Cultural Geography* 12:839–854.

deLind, Laura B. 2002. "Place, Work and Civic Agriculture: Common Fields for Cultivation." *Agriculture and Human Values* 19:217–224.

———. 2011. "Are Local Food and the Local Food Movement Taking Us Where We Want to Go? Or Are We Hitching Our Wagons to the Wrong Stars?" *Agriculture and Human Values* 28:273–283.

Downes, Lawrence. 2010. "In an Ugly Human-Trafficking Case, Hawai'i Forgets Itself." *New York Times*. September 21. http://www.nytimes.com/2010/09/21/opinion/21tue4.html?_r=0.

DuPuis, E. Melanie, and David Goodman. 2005. "Should We Go 'Home' to Eat? Toward a Reflexive Politics of Localism." *Journal of Rural Studies* 21:359–371.

Goodman, David. 2003. "The Quality 'Turn' and Alternative Food Practices: Reflections and Agenda." *Journal of Rural Studies* 19:1–7.

Guthman, Julie. 2004. *Agrarian Dreams: The Paradox of Organic Farming in California*. Berkeley: University of California Press.

Hamm, Michael W., and Anne C. Bellows. 2003. "Community Food Security and Nutrition Educators." *Journal of Nutrition Education and Behavior* 35:37–43.

Hassanein, Neva. 2003. "Practicing Food Democracy: A Pragmatic Politics of Transformation." *Journal of Rural Studies* 19:77–86.

Hawai'i Agricultural Statistics Service. 2012. "Seed Crop Industry Valued at New Record High $243 Million." *Hawai'i Seed Crops.* http://www.nass.usda.gov/Statistics_by_State/Hawaii/Publications/Sugarcane_and_Specialty_Crops/seed.pdf. Accessed August 13, 2013.

Hawai'i Department of Agriculture. 2003. *Agricultural Water Use and Development Plan 2003.* http://hdoa.Hawaii.gov/arm/files/2012/12/AWUDP-Dec2003.pdf. Accessed July 1, 2013.

Headey, Derek, and Fan Shenggen. 2008. "Anatomy of a Crisis: The Causes and Consequences of Surging Food Prices." *Agricultural Economics* 39:375–391.

Henderson, Janice Wald. 1994. *The New Cuisine of Hawai'i: Recipes from the Twelve Celebrated Chefs of Hawai'i Regional Cuisine.* New York: Villard Books.

Hitch, Thomas Kemper. 1992. *Islands in Transition: The Past, Present, and Future of Hawai'i's Economy.* Honolulu: First Hawaiian Bank and University of Hawai'i Press.

Ilbery, Brian, and Moya Kneafsey. 2000. "Producer Constructions of Quality in Regional Speciality Food Production: A Case Study from South West England." *Journal of Rural Studies* 16:217–230.

Johnston, Josee, Andrew Biro, and Norah MacKendrick. 2009. "Lost in the Supermarket: The Corporate-Organic Foodscape and the Struggle for Food Democracy." *Antipode* 41:509–532.

Kame'eleihiwa, Lilikalā. 1992. *Native Land and Foreign Desires: How Shall We Live in Harmony? Pehea Lā E Pono Ai?* Honolulu: Bishop Museum Press.

Kent, George. 2010. "Good Nutrition for All Is Better Goal than Food Self-Sufficiency." *Honolulu Star-Advertiser.* October 26. http://www.staradvertiser.com/editorial/good-nutrition-for-all-is-better-goal-than-food-self-sufficiency/.

Kloppenburg, Jack Jr., John Hendrickson, and G. W. Stevenson. 1996. "Coming in the Foodshed." *Agriculture and Human Values* 13:33–42.

Lang, Tim, and Michael Heasman. 2004. *Food Wars: The Global Battle for Mouths, Minds and Markets.* London: Earthscan/James & James.

Leung, PingSun, and Matthew Loke. 2008. *Economic Impacts of Increasing Hawai'i's Food Self-Sufficiency.* Honolulu: Cooperative Extension Service, College of Tropical Agriculture and Human Resources, University of Hawai'i at Mānoa.

Lobell, D. B., M. B. Burke, C. Tebaldi, M. D. Mastrandrea, W. P. Falcon, and R. L. Naylor. 2008. "Prioritizing Climate Change Adaptation Needs for Food Security in 2030." *Science* 319:607–610.

McAvoy, Audrey. 2014. "Why Hawaii Is Ground Zero for the GMO Debate." *Huffington Post.* April 21. http://www.huffingtonpost.com/2014/04/21/hawaii-gmo-flash-point_n_5187599.html.

McElroy, Jerome L., and Klaus de Albuquerque. 1990. "Sustainable Small-Scale Agriculture in Small Caribbean Islands." *Society and Natural Resources* 3:109–129.

Morgan, Kevin. 2010. "Local and Green, Global and Fair: The Ethical Foodscape and the Politics of Care." *Environment and Planning A* 42, no. 8:1852–1867.

Page, Christina, Lionel Bony, and Laura Schewel. 2007. "Island of Hawai'i Whole System Project." Boulder, CO and Kamuela, HI: Rocky Mountain Institute and the Kohala Center.

Pang, Gordon Y. K. 1998. "Waialua Seed Corn Plant Arouses Mixed Reactions." *Honolulu Star Bulletin.* August 21.

"Papaya Vandals Must Be Stopped." 2011. *Honolulu Star-Advertiser.* July 21. http://www.staradvertiser.com/editorial/papaya-vandals-must-be-stopped/.

Paulo, Walter Keli'iokekai, Duncan Ka'ohuoka'ala Seto, Eric M. Enos, Puanani Burgess, and Opelu Project Inc. 1999. *From Then to Now: A Manual for Doing Things Hawaiian Style.* Wai'anae, HI: Ka'ala Farm.

Paxton, Angela. 1994. *The Food Miles Report: The Dangers of Long-Distance Food Transport.* London: Sustainable Agriculture Food Environment (SAFE) Alliance.

Peters, C. W., Robert H. Reed, and C. Richard Creek. 1954. "Margins, Shrinkage and Pricing of Certain Fresh Vegetables in Honolulu." *Agricultural Economics Bulletin* 7.

Philipp, Perry F. 1953. *Diversified Agriculture of Hawai'i.* Honolulu: University of Hawai'i Press.

Raynolds, Laura T. 1997. "Restructuring National Agriculture, Agro-Food Trade, and Agrarian Livelihoods in the Caribbean." In *Globalizing Food: Agrarian Questions and Global Restructuring,* edited by D. Goodman and M. J. Watts, 119–132. London and New York: Routledge.

Sen, Amartya Kumar. 1981. *Poverty and Famines: An Essay on Entitlement and Deprivation.* Oxford and New York: Clarendon Press and Oxford University Press.

———. 1995. "Food Entitlements and Economic Chains." In *Hunger in History: Food Shortage, Poverty and Deprivation,* edited by L. F. Newman, 374–386. Oxford and Cambridge: Blackwell.

Statistics of Hawai'i Agriculture. Multiple Years. United States Department of Agriculture and Hawai'i Department of Agriculture. Hawai'i Field Office of the National Agricultural Statistics Service, 1428 S. King St, Honolulu, HI 96814.

Suryanata, Krisnawati. 2000. "Products from Paradise: The Social Construction of Hawai'i Crops." *Agriculture and Human Values* 17:181–189.

———. 2002. "Diversified Agriculture, Land Use and Agro-Food Networks in Hawai'i." *Economic Geography* 78:71–86.

Titcomb, Margaret, and Mary Kawena Pukui. 1951. "Memoir No. 29: Native Use of Fish in Hawai'i." *Journal of Polynesian Society* 60:1–146.

van der Ploeg, Jan Douwe. 2010. "The Food Crisis, Industrialized Farming and the Imperial Regime." *Journal of Agrarian Change* 10, no. 1:98–106.

Viotti, Vicki. 2010. "Robert Harris: The Leader of Sierra Club Hawai'i Wants a Marshall-Style Plan for the State that Would Lead to Greater Food Self-Sufficiency." *Honolulu Star-Advertiser.* October 22. http://www.staradvertiser.com/editorial/robert-harris/.

Welsh, Jennifer, and Rod MacRae. 1998. "Food Citizenship and Community Food Security: Lessons from Toronto, Canada." *Canadian Journal of Development Studies* 19:237–255.

Wilkins, Jennifer L. 2005. "Eating Right Here: Moving from Consumer to Food Citizen." *Agriculture and Human Values* 22:269–273.

1 | Tangled Roots

The Paradox of Important Agricultural Lands in Hawaiʻi

Krisnawati Suryanata and Kem Lowry

Concerns about Hawaiʻi's food and energy dependency have been voiced repeatedly. They tend to reach a peak whenever an economic disruption causes food or gas prices to go up. We hear oft-cited statistics that Hawaiʻi imports as high as 90 percent of its foodstuff, a reminder of the physical isolation of Hawaiʻi and our vulnerability to economic or natural disasters, shipping disruptions, and changes in energy costs. Many are also quick to point out with a tinge of nostalgia that in the not-too-distant past, Hawaiʻi was known as an agricultural state. Long before Hawaiʻi's economy became dependent on tourists, agriculture was the backbone of the economy.

In the mid-twentieth century, "agriculture" in Hawaiʻi meant export-oriented agriculture: sugar and pineapple production. At its peak, sugarcane and pineapple plantations occupied more than three hundred thousand acres of land in the state and directly and indirectly employed about one-third of the workforce. These two crops accounted for more than 90 percent of field crops (Hitch 1992). By the 1970s, however, high labor costs, foreign competition, and doubt about the continuation of federal price supports for sugar contributed to narrower profit margins and greater uncertainty. In addition, the state Supreme Court's McBryde decision in 1973 prohibited plantation owners from diverting water from one watershed to another, which had long been the accepted practice, thereby greatly increasing the costs of irrigation.

While acknowledging that plantation agriculture is not competitive in today's economy, many would raise the seemingly obvious questions: If Hawaiʻi's agricultural land resources could support the most vibrant sugarcane and pineapple production systems just four decades ago, why can't

Hawai'i's agricultural lands produce food for its urban populace? If the alternative agro-food networks (AAFNs) elsewhere thrive in spite of the pervasive power of globalization (e.g., Belasco 1989; Lappé and Lappé 2002), why can't Hawai'i's small farmers tap into the growing consumer awareness about the value of supporting local food systems? What can planners and policy makers do to ensure that the best lands remain available for local food production?

Food activists argue that alternative agro-food networks based on local production provide more than simple sustenance. They also represent efforts to restore ecological and social relationships in food production and consumption. AAFNs could provide a context for social action building towards a more sustainable, just, and democratic society (Kloppenburg, Hendrickson, and Stevenson 1996; Hendrickson and Heffernan 2002). While the ground for this optimism is real, a growing number of analysts caution against an uncritical celebration of localism in AAFNs (DuPuis and Goodman 2005; Born and Purcell 2006; Morgan 2010; deLind 2011). Promoting food democracy requires that we examine the degree to which alternative/local food systems address the objectives of social justice, ecological sustainability, and economic viability within a broader social movement that considers food as a human right.

In a place such as Hawai'i where access to agricultural lands has long been central to our economy and cultural identity, efforts to support local agriculture and food systems are an inevitable part of local politics. This chapter examines some of the key laws, regulations, and policy tools employed to manage agricultural and rural lands and how they have evolved over the last fifty years. These policies have been shaped over time by changes in international markets, demographic pressures, increasing local cultural awareness and practices, and the politics of land use. The result is a tangled, often contradictory, set of policies and programs intended to support multiple management goals for rural and agricultural lands.

Politics and Policies Affecting Agricultural Lands

Hawai'i's land-tenure pattern is partly the result of the transformation of the property rights system, called the Māhele (division), in 1848. This act led to the privatization and accumulation of land holding, primarily by non-Hawaiians who were well-situated to take advantage of the newly established property regime (Kameʻeleihiwa, this volume). By the late nineteenth century, a small group of merchants and sugarcane planters—Alexander and Baldwin, Castle and Cooke, American Factors, C. Brewer, and Theo H. Davies, collectively known as "the Big Five"—were able to dominate land tenure and use on the islands. The legacy of concentrated land ownership remains until today, with a relatively few large landowners controlling a large proportion of agricultural lands.

Statehood and the introduction of inexpensive jet travel triggered the rise of the tourism industry and an unprecedented urban and resort development boom (Callies 1984; Cooper and Daws 1985; Creighton 1978; Kent 1983). By this time, there were at least three major interest groups concerned about the fate of agricultural lands in Hawaiʻi. To the (plantation) agricultural industry, the lands were resources for commodity production and the backbone of the state's agricultural economy. To Native Hawaiians and subsistence farmers, the lands were livelihood and cultural resources critical for their survival. Last, to an emerging class of investors and developers, the lands were resources for urban expansion. Over the past half century, more interest groups with their respective visions for agricultural lands have emerged. They include the visitor industry that values agricultural lands for their aesthetic landscapes; the amenity migrants seeking to build their retirement or second homes on these lands; and a growing number of constituents who are concerned about ecological sustainability and local food security.

We identified three key regulatory mandates affecting agricultural lands, which summarized the political processes over the past half century (see table 1.1). First, the 1961 Land Use Law, which was enacted in response to concerns associated with the dramatic political and economic changes following statehood. The law was central to urban growth policies throughout the 1960s and 1970s, and it continues to frame subsequent policies associated with agricultural lands. Second, a 1978 Constitutional Amendment that explicitly states the importance of protecting agriculture passed by voters in the midst of a crisis affecting the plantation industry. Throughout the 1980s and 1990s this amendment helped justify the programs and policies intended to diversify agriculture and ameliorate the decline of traditional sugarcane and pineapple production. Third, the Important Agricultural Lands Laws (Act 183 and Act 233), passed in 2005 and 2008, respectively, that revived a mandate contained in the 1978 Constitutional Amendment to protect Important Agricultural Lands (IAL).

TABLE 1.1 Key regulatory mandates affecting agricultural lands

Time	*Key concerns triggering regulatory actions*	*Key regulatory mandate*
1961	Conversion of prime agricultural lands to urban use; scattered subdivisions on neighbor islands	The Land Use Law
1978	Crisis in plantation industry	Constitutional Amendment mandating the State to protect agriculture
2005, 2008	Proliferation of "fake farms"; escalating property values	Procedures to designate Important Agricultural Lands (Act 183/2005) and incentives (Act 233/2008)

These laws were passed in response to a growing concern that developers and speculators were exploiting the loopholes regarding allowable use in the Agricultural Districts, particularly as the real estate and second homes market was heating up since the late 1990s.

Throughout this period, concerns about food and energy security remain steady and have often been cited as key reasons to support different policies and programs affecting agricultural lands. However, as the following analysis will show, these goals remain peripheral to the political processes that drive the policies on agricultural lands.

The Land Use Law

Shortly following Hawai'i statehood in 1959, pressure to construct new subdivisions began to emerge on the island of O'ahu. Rapid expansion of the city of Honolulu began to consume prime agricultural lands in the central part of the island. There was also a scattered subdivision boom on the Big Island of Hawai'i. It became apparent that the planning and management capacities among the state's four counties to deal with the problems generated by the rapid urban development were uneven. The interests of the landowning oligarchy, state government, and agriculture labor unions converged and set the stage for the passage of the Hawai'i State Land Use Law

(LUL) in 1961, the first statewide zoning measure in the United States (Bossleman and Callies 1971; Callies 1984; Cooper and Daws 1985).

The Land Use Law (Hawai'i Revised Statutes §205) had two explicit objectives: to preserve prime agricultural land; and to prevent scattered, discontinuous patterns of urban settlement. The law established three major districts: Urban, Agriculture, and Conservation. A fourth district, Rural, was added in 1963 to accommodate pre-existing low-density rural residential areas on the Neighbor Islands. A State Land Use Commission, whose nine members are appointed by the governor and confirmed by the state senate, was charged with establishing boundaries among the four districts. As demarcated in 1964, all the built-up areas were designated as the Urban District, while forests and watershed protection area were assigned to the Conservation District. The Agricultural District—consisting of 52 percent of the land in the state—was a catch-all category that included lands used for cultivation of crops, as well as those not easily classified as urban or conservation. Even at the peak of agricultural economic dominance in the 1940s and 1950s, only a little over 300,000 acres out of the 2.1 million acres of land in the Agricultural District were under intensive crop cultivations (Kelly 1998). The rest was used for grazing, left idle, or intermittently cropped.

The Hawai'i State Land Use Law set forth a system of controls and incentives, including differential tax provisions that effectively reduce the taxes on agricultural lands to a mere fraction of what they would be if the land was reclassified for urban use. Landowners who dedicate their land for long-term agriculture also receive substantial reductions in land taxes. By the mid-1970s more than 420,000 acres were subject to tax dedication (Strong 1976). Large landowners were therefore not compelled to convert their holdings even when the lands were not actively cultivated.

The Land Use Commission's primary role is to act on petitions for boundary changes submitted by private landowners, developers, and state and county agencies. The commission's quasi-judicial decision-making process allows individual citizens to take part in the proceedings as "parties." Parties with direct interests may petition the commission to intervene in the proceeding. In addition to the petitioner, the Office of Planning and the respective County Planning Departments are mandatory parties to the proceedings. In an effort to streamline the decision-making process, the law was amended in 1985 to allow applicants for changes of land use of fifteen acres or less in Agriculture District to bypass the Land Use Commission and to apply directly to the counties.

The period from 1960 to 1980 was marked by the significant urban expansion on O'ahu, the closure of several plantations, and the increasing uncertainty about the future of agriculture. By the late 1970s, it was clear that state zoning and tax incentives had made an impact on the pace of conversion of agricultural lands. In spite of the immense pressure to develop and urbanize, the amount of agricultural lands lost to urban use was relatively low. In the twenty years following the enactment of the Land Use Law in 1964, a little over 40,000 acres of agricultural land were rezoned to urban use, amounting to less than 2 percent of the total agricultural lands (see table 1.2). However, the Land Use Law has not been as successful in preserving prime agricultural land. A study of urban redistricting petitions filed between 1964 and 1975 showed that in spite of the abundant underutilized agricultural land of poor quality, conversion to urban uses occurred largely on prime agricultural land on the island of O'ahu (Lowry 1980; Ferguson and Khan 1992; Santos-George et al. 1991). For example, in the late 1960s, major land owners Castle and Cooke's petition to develop what was eventually to become Mililani Town was approved by the Land Use Commission—setting the precedent for the gradual urbanization of hundreds of acres of prime agricultural land in Central O'ahu and Ewa. Criteria related to the location of new urban development, such as its proximity to employment and avoidance of scattered urban development, proved to be crucial variables in determining the land use policy of the Land Use Commission (Lowry 1980).

TABLE 1.2 Acreage by State Land Use districts, 1964–2013

	Acres in each State Land Use district				
Year	*Urban*	*Conservation*	*Agriculture*	*Rural*	*Total*
1964	117,800	1,862,600	2,124,400	6,700	4,111,500
1974	147,472	1,986,429	1,976,695	8,872	4,111,500
1984	158,620	1,969,351	1,974,236	10,181	4,112,388
1994	189,418	1,974,549	1,938,505	9,916	4,112,388
2004	196,991	1,973,636	1,931,378	10,383	4,112,388
2013	200,434	1,973,896	1,926,507	11,602	4,112,388

Source: State of Hawai'i Data Book 2013.

The 1978 Constitutional Amendment

State land use controls and lower tax rates for agricultural land proved to be insufficient in preventing the decline of plantation agriculture. Declining profits forced many plantations to close, leaving large former sugarcane and pineapple fields idle as their owners sought other uses. Landowners were struggling with how to make productive uses of agricultural land while state policy makers confronted the issues of alternative employment and housing for former plantation workers. In the context of this general anxiety about the future of agriculture, delegates to the 1978 constitutional convention approved an amendment to the Hawai'i State Constitution (Article XI, section 3) expressing its commitment to the protection of agriculture. The amendment requires the State "to conserve and protect agricultural lands, promote diversified agriculture, increase agricultural self-sufficiency and assure the availability of agriculturally suitable lands." It also requires the state legislature to develop standards and criteria to identify "Important Agricultural Lands" (IAL) that must be set aside and protected from conversion to urban uses.

Promoting Diversified Agriculture

For a time, diversified agriculture was viewed as the solution to the decline of plantation agriculture. The term "diversified agriculture" as originally conceived included all agricultural industries other than sugar and pineapple. Recognizing the vulnerability of Hawai'i's agricultural economy that depended upon these two export commodities, politicians and farmers alike began to search for new potentials in diversified agriculture at the end of World War II (Philipp 1953). When an increasing number of plantations closed down throughout the 1980s and 1990s, one of the initial strategies for developing Hawai'i's diversified agriculture sector was to continue the export-oriented strategy but diversify into high-value nontraditional export crops, such as tropical flowers, gourmet coffee, and tropical specialty fruits.

The strategy was modeled after the success in the marketing of pineapples in the early twentieth century (Dole and Porteus 1996), and of macadamia nuts in the 1970s. Both of these products developed a strong place-association with Hawai'i, and were marketed primarily to consumers in the mainland United States. The strategy, however, was vulnerable to the same structural problems that had diminished the traditional plantation industry. In a globalized market place, even products embedded with place-based uniqueness are increasingly incorporated into the mass commodity market, in which corporations with global networks of vertically integrated subsidiaries generally dominate. In these networks, Hawai'i growers have been marginalized and faced with declining prices in the global market (Suryanata 2000, 2002). By the mid-2000s it became clear that the "diversified" plantations growing coffee or specialty tropical fruits were not prospering as much as expected. The Waialua Estate Coffee grown on former sugar lands on O'ahu filed for bankruptcy in 2002 after operating for six years. The Guava Kai plantation, which had grown guavas on 500 acres of former sugar lands on the North Shore of Kaua'i, closed in 2007.

Niche market development tied to agro-tourism has fared slightly better. In this model, entrepreneurs depend on the hospitality industry and the sale of products marginally related to the farm, such as beverage coffee and tea; jams, jellies, and baked goods; and beauty products. A few of these agro-tourism enterprises have thrived, such as Ali'i Kula Lavender on Māui and the Hawaiian Vanilla Company on the Big Island of Hawai'i, but the extent to which the industry can expand remains uncertain.

Another strategy of agricultural diversification calls for increasing production for local consumption. Hawai'i's dependency on imported foods had prevailed even during the heyday of the plantation economy. This dependency only intensified as the state's economy developed. Because of the isolated geographic location and small population, growers of generic food products in Hawai'i face a "pocket market" problem, a situation in which local producers cannot effectively compete outside the region (Peters, Reed, and Creek 1954). Thus, most of Hawai'i-grown vegetables have no competitive edge outside Hawai'i, and virtually all are destined for consumption within the state. Growers do not really have the option of selling their products outside Hawai'i when supplies are large enough to depress local prices. The spatial distribution of Hawai'i's population further complicates the development of diversified agriculture for local consumption. About 70 percent of Hawai'i's 1.42 million residents live in the city and county of Honolulu on the island of O'ahu. Because of shipping costs, neighbor island producers do not hold much competitive advantage over producers from the mainland or abroad.

The increased globalization of agro-food systems has allowed Hawai'i's consumers to enjoy the variety of products offered by producers from all over the world. Vertically integrated food companies, with their extensive networks of growers, processors, transportation agents, and retailers, efficiently deliver food products across vast distances. While the relatively low prices of imported food benefit consumers in general, they also increase the level of competition for Hawai'i's producers. Even in the specialty market of organic produce, island producers have found that their small but growing market is quickly dominated by the conventional produce delivery systems from the mainland United States (Suryanata 2002).

Analysis of the different strategies of diversified agriculture shows the enormous challenge for Hawai'i's agriculture to position itself within the globalized agro-food systems. In the export-oriented strategy, the markets for "specialty" products that used to be associated with Hawai'i—such as pineapples, macadamia nuts, or tropical fruits and flowers—are increasingly incorporated into the global mass commodity market in which Hawai'i growers do not hold competitive advantage. Meanwhile, the import-substitution strategy has also been facing an uphill battle. Despite the efforts to promote production for local consumption, Hawai'i's food dependency has remained relatively unchanged in the food economy of the islands (Southichack 2007).

Identifying Important Agricultural Lands

In addressing the constitutional mandate to protect important agricultural lands, the initial emphasis was technical. The Land Use Commission had been relying on agricultural productivity ratings developed by the Land Study Bureau (LSB) of the University of Hawai'i to guide their rulings. In a series of reports published between 1965 and 1972, the LSB rated agricultural lands as A, B, C, D, and E in terms of their overall productivity. The LSB rated 203,244 acres of agricultural land as Class A or B, which are generally classified as "prime" agricultural lands (State Agriculture Plan, Technical Reference Document, October 1982, II-91).

In 1977, staff from the College of Tropical Agriculture of the University of Hawai'i and the state Department of Agriculture started adapting a land rating system of the US Department of Agriculture to create a classification system called "Agricultural Lands of Importance to the State of Hawai'i" (ALISH). The ALISH system includes three classes of important agricultural lands: prime (soils with best physical, chemical, and climatic properties for mechanized field crops); unique (land other than prime for unique high-value crops such as coffee, taro, and watercress); and other important lands (nonprime, nonunique, but important lands requiring irrigation or commercial production management). The ALISH system, adopted by the Department of Agriculture in 1977, resulted in 978,174 acres being classified as prime, unique or other (State Agriculture Plan, Technical Reference Document, October, 1982, II-94). Of this amount, 335,630 acres were classified prime or unique.

In 1983, the legislature initiated another land classification system. Act 273 established the State Land Evaluation and Site Assessment (LESA) Commission charged with establishing standards and criteria for identifying and inventorying important agricultural lands. The resultant LESA numeric rating system combined land evaluation (soils, topography, climate) and site assessment (nonphysical properties including location and land use). The LESA rating system resulted in a designation of about 760,000 acres of agricultural land—or about 41 percent of the lands in the Agriculture District—as "important" (State of Hawaiʻi Land Evaluation and Site Assessment Commission 1986). However, like the LSB and ALISH before it, the LESA system was not adopted by the state legislature. By then, the support for LESA among state and county officers had diminished because it no longer seemed relevant. Several basic premises in developing the LESA rating system to identify important agricultural lands, such as the prime importance of the sugar and pineapple industries, were no longer valid in light of the rapid decline of those industries.

The efforts to develop a method for designating Important Agricultural Lands coincided with several other efforts to weaken the Land Use Commission's role in governing changes in land use. County officials argued that the conditions that prevailed in the early 1960s—a strong state government and relatively weak county governments—had greatly changed. They have argued for more "home rule" over important land use and environmental decisions. More importantly, key constituents who had supported the 1978 constitutional amendment began to question what "protection" might mean in terms of a new or more restrictive regulatory regime. At a time when substantial amounts of former plantation land were becoming available for alternative uses, many landowners feared that new regulations would greatly limit their land use options. Advocates of small farms, on the other hand, feared that the marginal lands where many of these farms are located might be excluded from the IAL classification, and therefore left "unprotected." Such exclusion would deny them any benefits extended to agricultural lands that have thus far been critical in keeping the farms afloat.

The impasse over the technical criteria to identify IAL left Hawai'i's agricultural land policy in limbo for more than two decades. The Land Use Law—imperfect as it was—continued to be the primary regulatory tool to manage agricultural lands. While the Land Use Law guided urban development at the urban fringe, it did little to manage the de facto use of lands in the Agricultural District. Following the law's stipulation on allowable use in the Agricultural District, numerous agricultural subdivisions with nominal or no agricultural activities—dubbed as "fake farms"—were built as residential development in all the rural counties.

Important Agricultural Lands (Act 183 and Act 233)

Small-scale gentleman farms became a standard land use throughout the Agricultural District. However, the practice had never been scrutinized until a major controversy emerged involving Hokuli'a, a $1 billion project to build an exclusive golf resort on 1,550 acres of agricultural land in West Hawai'i. The developer had acquired a permit from the County of Hawai'i in 1998 under a 1976 provision in the Land Use Law that allows conversion of agricultural land as long as the development included some farming component. In 2000, opponents sued the developer to stop the project, claiming that the luxury development failed to get approval from the State Land Use Commission for what is essentially a suburban-type development in the state-designated Agriculture District. In September 2003, Hawai'i Circuit Judge Ronald Ibarra ordered the project to halt construction (Carlton 2005). The court ruled that it was illegal to build luxury homes on land zoned for agriculture. By the time the project was shut down, the developer had spent $300 million and more than 150 lots had been sold. The ruling raised many questions about the legal status of thousands of homes that that had been built on agricultural lands, the county's authority to manage growth outside the urban district, and more generally, the direction of rural change in the face of declining agriculture and increased global demand for housing in paradise.

While it would be easy to attribute the fake farms controversy to rent-seeking behavior of major landowners who take advantage of loopholes in the Land Use Law, the problem runs much deeper and reflects the challenges facing rural regions experiencing counter-urbanization and amenity migration (Boyle, Halfacree, and Robinson 1998; Duane 1999; Loeffler and Steinicke 2007). The real estate boom from the late 1990s to 2008 amplified the impacts of this worldwide trend. Second-home purchases, a growing trend throughout the nation, grew even faster in Hawai'i. As Hall and Müller (2004) argue, second homes are an integral part of contemporary tourism. While real estate transactions in the US mainland began to slump in 2006, Hawai'i's median home prices were more stable and did not experience a similar decline. The weakened US dollar against foreign currencies also made Hawai'i's real estate an attractive investment opportunity for foreign buyers (Schaefers 2007). To illustrate this hypermarket, 37 percent of all Māui resident housing sales in 2004 were to buyers residing outside the county. In some districts, the number of offshore sales reached a staggering 52 percent of total sales (Māui County Planning Department 2006). In some places, this influx of amenity migrants (cf., Moss 2005) has resulted in increased traffic congestion and demands for other public services and facilities. Growing resentment over the rising cost of housing, the direction of rural transformation, and the process that was perceived to favor the super-rich eventually contributed to the Hokuli'a controversy mentioned above.

In 2001, a legislative committee convened an Agriculture Working Group (AWG) composed of state and county agency staff, farm organization representatives, private landowners, conservationists, and others to address how best to respond to the constitutional mandate and to develop draft legislation for the 2004 legislative session. After meeting monthly for three years, the AWG produced the draft legislation that provided criteria for designating important agricultural lands. The AWG draft also recommended that incentives such as grants, preferential tax assessment, or funding mechanisms for agricultural preservation be provided to the landowners whose lands are designated as "important agricultural lands" (Suarez 2005).

The report of the Agricultural Working Group was presented to the legislature in the context of this politically charged atmosphere. Legislators and state and county officials were eager to rectify what they viewed as a major fault in the regulatory structure that governed the rapidly changing rural areas. In 2005, twenty-seven years after the constitutional mandate to protect important agricultural lands, Act 183 was signed into law to establish standards, criteria, and mechanisms to identify Important Agricultural Lands and to develop policies for incentives to encourage long-term retention of important agricultural lands for agricultural use.

Following Act 183, Important Agricultural Lands can be designated in two ways. A land owner may file a petition seeking designation of the owner's land as "important agricultural land." The State Land Use Commission would then review the petition for consistency with important agricultural land policies and criteria, and approve the petition, if two-thirds of the members of the commission concur. In addition, the counties can also identify lands for potential designation, consistent with community recommendations and existing plans such as general plans and community plans. Subsequently, legislators passed Act 233 in 2008 to stipulate incentives to encourage landowners' participation in designating the IAL. The new incentives include up to $7.5 million for costs such as roads, utilities, wells, reservoirs, agricultural housing, and technical studies of various kinds; a loan guaranty to commercial lenders to facilitate IAL borrowers' eligibility to low-interest loans; a fast-track approval process for the designation of IAL and an 85/15 provision, allowing a landowner to petition to the Land Use Commission to reclassify up to 15 percent of important agricultural land units into Conservation, Rural, or Urban District. For an agricultural land unit of several thousand acres, the amount of land potentially eligible for urban development could be substantial.

Thus far, only the landowner-initiated process has been used to designate important agricultural lands. Alexander and Baldwin—one of the Big Five—designated 27,000 acres on Māui and 3,700 acres on Kaua'i (Gomes 2011a). Other large landowners, including Molokai Properties Ltd. on Molokai and Castle and Cooke on O'ahu, have also petitioned to designate part of their lands as Important Agricultural Lands (Gomes 2011b)—but all have claimed that they would waive the right to reclassify 15 percent of their lands. Unfortunately, the voluntary designation by large landowners would neither address the structural problems faced by Hawai'i's agriculture nor those faced by county and state planners confronting rural gentrification.

On rural gentrification, Daniels (1997, 2000) argues that where farming economies are strong, comprehensive farmland preservation programs can be an effective rural policy that not only supports the agricultural sector, but also maintains open space and prevents sprawling that can drive up the cost of public services. Examples of successful farmland preservation include Lancaster County in Pennsylvania and Montgomery County in Maryland. However, in places such as Hawai'i where the agricultural economy is relatively weak, the best outcome one can expect from such programs is the retention of some kind of "rural character" featuring open space. The other two objectives often used to justify such a policy—the preservation of agriculture and the reduction of sprawling—would remain elusive. Daniels further argues that preferential tax treatment would more likely subsidize a rural lifestyle than foster a viable agricultural economy, and that "maintaining a working landscape and maintaining open space require different planning strategies and land use techniques" (2000, 262).

Likewise in Hawai'i, the protection of important agricultural land has had little impact on strengthening farm production. Meanwhile, the proliferation of vacation homes and gentleman farms on agricultural lands continues, bringing further concerns such as escalating property values and alienation of the local community. A 1980 provision of the Land Use Law allows a county to permit other uses in the state-designated Agricultural District if the land in concern is less than fifteen acres in size (figure 1.1).

FIGURE 1.1. Important agricultural land in South Kohala, Hawai'i. Photo by Krisna Suryanata.

This provision has facilitated development in a piecemeal fashion involving transactions between small landowners and individual buyers. Some state officials have noted that this more decentralized rural change has affected rural Hawai'i more than the high-profile large-scale developments that often draw well-organized opposition.

The Paradox of Important Agricultural Lands

The identification and designation of important agricultural land seems to have become an end in itself rather than a means to facilitate the achievement of a vision for a new agricultural future for Hawai'i. It has had as little impact on the processes of rural gentrification as it has had on improving food security for the islands. Much of the second-home developments in the Agricultural District continue to occur without any planning guidance. Meanwhile, efforts to revive Hawai'i's agricultural sector continue to face an uphill battle. By 2012, roughly 5 percent (less than 100,000 acres) of the 1.9 million acres of agricultural land are under cultivation (*State of Hawai'i Data Book* 2013). The rest has been left idle, used as grazing land, or intermittently cropped.

It is clear that the challenges facing Hawai'i's agriculture are not limited to the availability of prime agricultural lands. Farmers also cite access to water, capital, and labor as the barriers to expanding production (Park 2010; Bondera 2011), as well as the need to develop business capacities among farm operators. The State Food Safety Program designed to protect consumers has inadvertently affected local small producers who do not have the capacity to meet the certification standard. Because major grocery retailers and distributors in the state use the standard, many of these farms are excluded from the largest portion of consumer market (Santoro 2011). Changes in the program that ensure clear guidelines and prompt inspection could reduce this barrier.

A resilient local food production system requires a sufficient supply of local products *and* a sustained demand by the majority of its urban populace. This requires public policies and private investments to go beyond merely protecting the farms to also improving access to quality local food for the general public. At present, the few successful farms producing for local consumption are those that have capitalized on small-scale initiatives including direct marketing, farm-chef collaborations, and producer cooperatives. These initiatives connect producers with select groups of consumers who support local food productions *and* can afford the extra time and/or costs attached to the local products. While this niche market orientation might be an effective business strategy for individual producers, it cannot be the basis of strategy to ensure food security.

More broadly, our understanding of food security must go beyond the dichotomy of locally grown vs. imported food. Hawai'i residents also have the tastes, appetites, and budgets that are likely to keep them integrated in the global food networks. For better or worse, their appetites for the enormous variety of processed foods, for fruits and vegetables grown more cheaply elsewhere, for Japanese *unagi* (freshwater eel), French cheese, and New Zealand wines will continue to drive their patterns of consumption. Ensuring food security through increasing local production is therefore not an uncontested goal, especially when it competes with other visions about how Hawai'i's agricultural and rural lands should be used.

In focusing so exclusively on prime agricultural land preservation, our attention is diverted from the ultimate need to reconcile the many visions for the future of Hawai'i's agricultural and rural lands. Some argue for developing diversified agriculture industry that not only includes edible fruits and vegetables, but also biofuels, seed crops (see Schrager and Suryanata, this volume), flowers, and nursery crops. Others view the potential of agricultural lands for agro-tourism, landscape protection, and renewable energy production. While these lands are designated as "agricultural lands," they also attract those who seek to build villas or hideaways in rural settings. Still others seek lower cost housing on agricultural land without the site development requirements (and expense) of urban subdivision codes.

Crafting effective rural policy requires us to recognize the multiple visions for agricultural lands, along with the political stakeholders who advocate these visions. The globalized food system, international and national investments in amenity housing, and the local politics of land use will continue to shape our rural and agricultural policy making. Reconciling urban uses, alternative types of agriculture, and landscape protection will require more nuanced land management policy than is evident in the current efforts to preserve some high-valued agricultural lands.

References

Allen, Patricia, et al. 2003. "Shifting Plates in the Agrifood Landscape: The Tectonics of Alternative Agrifood Initiatives in California." *Journal of Rural Studies* 19, no. 1:61–75.

Belasco, Warren James. 1989. *Appetite for Change: How the Counterculture Took on the Food Industry, 1966–1988*. New York: Pantheon Books.

Bondera, Melanie. 2011. "From Farm to Fork." *Honolulu Weekly*. June 8. http://honoluluweekly.com/feature/2011/06/from-farm-to-fork/.

Born, Brandon, and Mark Purcell. 2006. "Avoiding the Local Trap: Scale and Food Systems in Planning Research." *Journal of Planning Education and Research* 26:195–207.

Bossleman, Fred, and David Callies. 1971. *The Quiet Revolution in Land Use Control*. Washington, DC: US Government Printing Office.

Boyle, P. J., Keith Halfacree, and Vaughan Robinson. 1998. *Exploring Contemporary Migration*. London and New York: Longman.

Callies, David. 1984. *Regulating Paradise: Land Use Controls in Hawai'i*. Honolulu: University of Hawai'i Press.

Carlton, Jim. 2005. "Land-Use Ruling Shakes Hawai'i Developers." *Wall Street Journal*. July 6, B.1.

Cooper, George, and Gavan Daws. 1985. *Land and Power in Hawai'i: The Democratic Years*. Honolulu: Benchmark Books.

Creighton, Thomas Hawk. 1978. *The Lands of Hawai'i: Their Use and Misuse*. Honolulu: University of Hawai'i Press.

Daniels, Thomas L. 1997. "Where Does Cluster Zoning Fit in Farmland Protection?" *Journal of the American Planning Association* 63, no. 1:129–137.

———. 2000. "Integrated Working Landscape Protection: The Case of Lancaster County, Pennsylvania." *Society & Natural Resources* 13, no. 3:261–271.

deLind, Laura B. 2011. "Are Local Food and the Local Food Movement Taking Us Where We Want to Go? Or Are We Hitching Our Wagons to the Wrong Stars?" *Agriculture and Human Values* 28:273–283.

Dole, Richard, and Elizabeth Dole Porteus. 1996. *The Story of James Dole*. Honolulu: Island Heritage Publishing.

Duane, Timothy P. 1999. *Shaping the Sierra: Nature, Culture, and Conflict in the Changing West*. Berkeley: University of California Press.

DuPuis, E. Melanie, and David Goodman. 2005. "Should We Go 'Home' to Eat? Toward a Reflexive Politics of Localism." *Journal of Rural Studies* 21, no. 3:359–371.

Ferguson, Carol, and M. Akram Khan. 1992. "Protecting Farmland Near Cities: Trade-Offs with Affordable Housing in Hawai'i." *Land Use Policy* 9 (October): 259–271.

Gomes, Andrew. 2011a. "Molokai Ranch Owners Seek Ruling on Land Use." *Honolulu Star-Advertiser*. January 4. http://www.staradvertiser.com/business/molokai-ranch-owners-seek-ruling-on-land-use/.

———. 2011b. "Castle & Cooke Filing Could Help Farmers." *Honolulu Star-Advertiser*. January 5. http://www.staradvertiser.com/business/castle-cooke-filing-could-help-farmers/.

Hall, C. Michael, and Dieter Müller, eds. 2004. *Tourism, Mobility and Second Homes: Between Elite Landscape and Common Ground.* Clevedon: Channelview Publications.

Hendrickson, M. K., and W. D. Heffernan. 2002. "Opening Spaces through Relocalization: Locating Potential Resistance in the Weaknesses of the Global Food System." *Sociologia Ruralis* 42, no. 4:347–369.

Hitch, Thomas Kemper. 1992. *Islands in Transition: The Past, Present, and Future of Hawai'i's Economy.* Honolulu: First Hawaiian Bank and University of Hawai'i Press.

Kame'eleihiwa, Lilikalā. 1992. *Native Land and Foreign Desire: Pehea Lā E Pono Ai?* Honolulu: Bishop Museum Press.

Kelly, James L. 1998. "Agriculture." In *Atlas of Hawai'i,* edited by J. O. Juvik and S. P. Juvik, 246–251. Honolulu: University of Hawai'i Press.

Kent, Noel. 1983. *Hawai'i: Island under the Influence.* New York: Monthly Review Press.

Kloppenburg, Jack Jr., John Hendrickson, and G. W. Stevenson. 1996. "Coming in the Foodshed." *Agriculture and Human Values* 13, no. 3:33–42.

Lappé, Frances Moore, and Anna Lappé. 2002. *Hope's Edge: The Next Diet for a Small Planet.* New York: Jeremy P. Tarcher/Putnam.

Loeffler, Roland, and Ernst Steinicke. 2007. "Amenity Migration in the US Sierra Nevada." *Geographical Review* 97, no. 1:67–88.

Lowry, G. Kem. 1980. "Evaluating State Land Use Control: Perspectives and Hawai'i Case Study. *Urban Law Annual* 18:85–127.

Māui County Planning Department. 2006. *Socio-economic Forecast: The Economic Projections for the Māui County General Plan 2030.* Wailuku: County of Māui.

McBryde Sugar Company Limited v. Aylmer F. Robinson et al. 1973. No. 4879, Supreme Court of Hawai'i, December 20.

Morgan, Kevin. 2010. "Local and Green, Global and Fair: The Ethical Foodscape and the Politics of Care." *Environment and Planning A* 42, no. 8:1852–1867.

Moss, Laurence A. G., ed. 2005. *The Amenity Migrants: Seeking and Sustaining Mountains and Their Cultures.* Wallingford: CAB International.

Park, Gene. 2010. "Local Farms in Labor Bind." *Honolulu Star-Advertiser.* September 13. http://www.staradvertiser.com/hawaii-news/local-farms-in-labor-bind/.

Peters, C. W., Robert H. Reed, and C. Richard Creek. 1954. "Margins, Shrinkage and Pricing of Certain Fresh Vegetables in Honolulu." *Agricultural Economics Bulletin* 7.

Philipp, Perry F. 1953. *Diversified Agriculture of Hawai'i.* Honolulu: University of Hawai'i Press.

Santoro, Al. 2011. "Food Solutions Nobody's Talking About." *Honolulu Weekly.* July 27. http://honoluluweekly.com/cover/2011/07/food-solutions-nobodys-talking-about/.

Santos-George, A., et al. 1991. "Urban Conversion of Hawai'i's Agricultural Lands, 1975–89." *Journal for Hawaiian and Pacific Agriculture* 3:1–8.

Schaefers, Allison. 2007. "It's Busy at the Top in the Hawai'i Luxury-Home Market." *New York Times.* December 7. http://travel.nytimes.com/2007/12/07/travel/escapes/07hawaii.html.

Southichack, Mana. 2007. "Inshipment Trend and Its Implications on Hawai'i's Food Security." Honolulu: Hawai'i Department of Agriculture.

State of Hawai'i Data Book. 2013. Honolulu: Department of Business, Economic Development and Tourism, Research and Economic Analysis Division, Statistics and Data Support Branch. http://dbedt.hawaii.gov/economic/databook/db2013/. Accessed January 16, 2014.

State of Hawai'i Land Evaluation and Site Assessment Commission. 1986. *A Report on the State of Hawai'i Land Evaluation and Site Assessment System*. Honolulu: Legislative Reference Bureau.

Strong, Ann L. 1976. "Deferred Taxation—Long Rollback." In *Untaxing Open Space: An Evaluation of the Effectiveness of Differential Assessment of Farms and Open Space*, edited by J. C. Keene, D. Barry, R. Coughlin, J. Farnam, E. Kelly, T. Plaut, and A. L. Strong, 164–201. Philadelphia: Regional Science Research Institute.

Suarez, Adrienne Iwamoto. 2005. "Avoiding the Next Hokuli'a: The Debate over Hawai'i's Agricultural Subdivisions." *University of Hawai'i Law Review* 27:441–468.

Suryanata, Krisnawati. 2000. "Products from Paradise: The Social Construction of Hawai'i Crops." *Agriculture and Human Values* 17, no. 2:181–189.

———. 2002. "Diversified Agriculture, Land Use and Agro-Food Networks in Hawai'i." *Economic Geography* 78, no. 1:71–86.

2 | Food Security in Hawai'i

George Kent

According to the Food and Agriculture Organization (FAO) of the United Nations, "Food security exists when all people, at all times, have physical, social and economic access to sufficient, safe and nutritious food to meet their dietary needs for an active and healthy life" (FAO 2009, 8). Food insecurity can take many different forms. This chapter explores three broad concerns for Hawai'i: overall food supply, disasters, and poverty. Each of these broad categories covers a variety of specific issues. For example, overall food supply is about food quantity and quality now and in the future, under various contingencies. It would include considerations of agriculture, processing, transport, infant feeding, nutrition-related health problems, genetically modified organisms, and related issues, under various long-term economic and climate scenarios. Disaster refers not only to tsunamis and earthquakes but also to economic collapse, terrorism, food supply crises, and other kinds of emergencies. Poverty refers to difficulties in obtaining adequate food by various categories of low-income individuals, and also the status of the economy as a whole. Thus, food security must be recognized as multidimensional, raising a broad variety of concerns for which policies and planning are needed. Also, Hawai'i must learn to differentiate between food security, self-sufficiency, and resilience, which are related, but not the same.

Overall Food Supply

Before contact, Hawai'i was self-sufficient in terms of food, by necessity, but not by choice. There were periodic famines (wī), usually due to disruptive events such as epidemics and warfare (Schmitt 1970). After contact, Hawai'i became involved in exporting, sometimes with serious consequences:

> Because the chiefs and commoners in large numbers went out cutting and carrying sandalwood, famine was experienced from Hawaii to Kauai . . . The people were forced to eat herbs and fern trunks, because there was no food to be had. When Kamehameha saw that the country was in the grip of a severe famine, he ordered the chiefs and commoners not to devote all their time to cutting sandalwood. (Kuykendall, as quoted in Schmitt 1970, 113)

If the efforts used to harvest sandalwood were instead devoted to harvesting or raising food, the famine could have been averted.

Hawai'i later devoted much of its land to producing and exporting food products. In the 1860s a new variety of rice was introduced, rapidly replacing taro. In 1862 Hawai'i exported more than 100,000 pounds of milled rice to California, and more than 800,000 pounds of paddy (Haraguchi 1987). Much of this production was made by Chinese and Japanese immigrant laborers. Later, sugar and pineapple became the dominant export crops. Hawai'i was for a time a major exporter of food products, based on its sugar and pineapple production, but that era is over.

Much of Hawai'i's food production for local consumption has now been displaced by imported food. This has raised alarm in many quarters. In 2012 the self-sufficiency bill of the Hawai'i State House, HB2703 HD2, said Hawai'i is dangerously dependent on imported food:

> As the most geographically isolated state in the country, Hawaii imports approximately ninety-two percent of its food, according to the United States Department of Agriculture. Currently, Hawaii has a supply of fresh produce for no more than ten days. Ninety percent of the beef, sixty-seven percent of the fresh vegetables, sixty-five percent of the fresh fruits, and eighty percent of all milk purchased in the State are imported. The legislature further finds that Hawaii's reliance on out-of-state sources of food places residents directly at risk of food shortages in the event of natural disasters, economic disruption, and other external factors beyond the State's control. (Hawaii State Legislature 2012)

Most analysts agree that Hawai'i currently imports 85 percent or more of its food from the US mainland and other countries (Leung and Loke 2008; Page et al. 2007). Some analyses focus specifically on imports and local production of fresh fruits and vegetables (Lee and Bittenbender 2007; Southichack 2007).

The self-sufficiency bill raised the import-replacement argument.

> The legislature further finds that each food product imported to Hawaii is a lost opportunity for local economic growth. The legislature notes that according to the University of Hawaii College of Tropical Agriculture and Human Resources, an increase in the production and sale of Hawaii-grown agricultural commodities would contribute to significant job creation. The research shows that replacing ten percent of current food imports will create a total of two thousand three hundred jobs. (Hawaii State Legislature 2012)

However, this analysis favors the producers' perspective and does not give sufficient attention to the consumers' perspective. Increasing purchases of locally produced foods would benefit local farmers, but it could also mean that consumers have to pay higher prices. The main reason Hawai'i imports much of its food is that it cannot produce the food as cheaply as it can import it.

Many people believe that the long-distance transportation of food to Hawai'i leads to high economic and environmental costs. However, there has not been any broad study of these impacts. Ocean transport is relatively cheap, in terms of unit cost per mile. Its environmental impacts may be comparable to that of ground-based transport. Food prices in Hawai'i's stores may be due not so much to transport costs as to the fact that the retailers face less competition in Hawai'i than they do on the mainland, and thus can charge higher prices. Higher real estate and utility costs also have to be considered. Oceanic transport costs are not as high as many people assume. It is true that the Jones Act,[1] which requires that shipping between US ports must be done on US-flagged ships, increases transport costs, but quantitative estimates of the Jones Act's impact on food prices are not available.

Food security in Hawai'i is often understood in terms of possible interruptions to food imports, but there are other possible threats as well. For example, things like local climate change, bee mites, and disruptions in water supply could threaten Hawai'i's agriculture. Local economic weaknesses of various kinds can lead to sharp reductions in local food production, as we have seen in dairy and meat production. There are also dangers that can arise at the consumer end of the food system. Hawai'i has been fortunate so far in not having had any major food safety incidents, but there are safety risks. Hawai'i relies mainly on the federal government to ensure food safety. The state Department of Health (HDOH) inspects restaurants and food markets only infrequently. Others are concerned about the impacts of genetic modification of food products, because a large portion of Hawai'i's agricultural land is devoted to the research on seeds for genetically modified products (Conrow 2009). They are concerned by the economic and environmental impacts at the production end and the health impacts at the consumption end.

Expand Food Production?

Increasing food production for local consumption faces numerous challenges. First, competing uses of the land such as encroaching housing developments and golf courses take land out of agriculture. Second, much of the agricultural land is used to produce crops other than food, such as seeds, biofuels, and ornamentals. Third, where there is food production, much of it is not basic food. Items such as coffee, macadamia nuts, and herbs would not be needed when food supplies are short. Fourth, some of the food that is produced is exported from the state. Fifth, it can be difficult for local farmers to compete with producers elsewhere who face lower land and labor costs.

Noncommercial food production could be increased in Hawai'i by making better uses of lawns, rooftops, and schoolyards. The worldwide movement to promote urban agriculture has developed many techniques that could be promoted in Hawai'i. The LEAF Project has several demonstration projects already under way, as shown at http://leafhawaii.org.

In some cases, the promotion of agriculture is mainly about protecting the livelihoods of small farmers, not about the products they deliver. For example, in the struggle to preserve the small farms in Kamilo Nui Valley in Hawai'i Kai, its defenders have not claimed that this valley has been making an important contribution to the state's food supply. It is important as the basis for the livelihood of the farmers who work the land there. Similarly, while the front page of the local newspaper may headline, "Blight Threatens Basil" (Nakaso 2011), it evokes little concern about Hawai'i's basic food supply. The objectives of ensuring food security and protecting farmers' livelihoods are both important, but they should not be confused with one another.

Hawaiʻi's supply of land is limited, but its supply of ocean is not. However, food production in the ocean is difficult. Natural marine fisheries around the islands have never been highly productive because of the great depth of near shore waters, and the absence of nutrient upwelling associated with continental shelves. The reef fisheries have been severely depleted, so the great majority of fish consumption in Hawaiʻi is based on imports. There are attempts to revive traditional aquaculture methods, but they do not produce large volumes. Modern commercial aquaculture in Hawaiʻi has a checkered business record, with highly publicized ambitious start-ups often followed by quiet shutdowns. Some of the operations are owned and operated by businesses based outside Hawaiʻi, and produce primarily for export, thus contributing little to the local food supply. There is evolving interest in aquaponics as an environmentally friendly method of combining aquaculture and farming.

Expanding agriculture would have a limited impact on Hawaiʻi's food system. Hawaiʻi's farm revenue in 2009 totaled $629 million, just short of the record of $642 million set in 1980 (Gomes 2011; also see USDA 2011). Much of the farm revenue is from nonfood products for exports such as seeds and ornamentals. Hawaiʻi's farm revenue attributable to food consumed within the state is about $400 million per year. Hawaiʻi's total food imports are roughly $2 billion per year. On this basis, Hawaiʻi farms produce roughly 20 percent of the state's food supply, in terms of monetary value. Probably about 80 percent of the imports are from the US mainland, and the remaining 20 percent from the rest of the world. A substantial share of the food produced and consumed in Hawaiʻi goes to military families and tourists. Perhaps that share should be excluded from calculations about the degree to which local agriculture contributes to local food self-sufficiency. If local food production operations are owned by outsiders, the profits would go elsewhere, and it is not clear that these operations really contribute to local self-sufficiency. Many of Hawaiʻi's agriculture workers have been immigrants, and that pattern is likely to continue in the future. The implications of this production structure for Hawaiʻi's self-sufficiency should be given some thought.

As in other developed economies, most of Hawaiʻi's food money goes to processors, not farmers, especially the processors outside Hawaiʻi. Only about 7,300 people work in food processing in Hawaiʻi (Yonan 2011). There are opportunities to expand the food processing sector, but the potential is limited because Hawai'i's processors must work with high costs and small volumes, and compete with large-scale processors based elsewhere (Hawai'i Food Manufacturers Association 2011).

Critics have argued that the arguments for localization may be overstated (Dean 2007; Singer and Mason 2007; Desrochers and Shimizu 2008; DeWeert 2009; McWilliams 2007; Roberts 2009). In Hawai'i, pushing food self-sufficiency too far or in the wrong way could increase costs for consumers, and it could reduce local food security by creating overdependence on one source. If it is not managed well, it could lead to the depletion of local resources. Increasing self-sufficiency could be advantageous to certain groups in the state, such as farmers, while being disadvantageous to others, such as the nonfarming poor.

There is a great deal of enthusiasm for increasing local food self-sufficiency, in the state government and in the community (to illustrate, see http://hawaiihomegrown.net/). However, there is a need for discussion about how far and how fast it should go. The degree of self-sufficiency is not something that should be maximized. It should be optimized, taking a broad variety of considerations into account. While it is true that Hawai'i's physical environment could support the production of a wide variety of food products, the high land and labor costs make it difficult to increase production. Moreover, Hawai'i could not produce the wide variety of products it now imports. Achieving full self-sufficiency would be impractical, and getting close to full self-sufficiency would require radical changes in people's diet and lifestyle. Although some people might welcome those changes, others would not. In short, it would be good for Hawai'i to have the *capacity* to be food self-sufficient just in case it was suddenly isolated from the rest of the world. But if Hawai'i pushes for *actual* self-sufficiency long before it is needed, its people would forgo the benefits that come with trade. It would be like moving the family into the basement now because a storm is likely to come in the next few years. Preparing is one thing; doing it is something else.

Resiliency

The 2012 food self-sufficiency bill said, "increasing local production will ensure that Hawai'i's food sources will be more resilient to global supply disruptions, better able to cope with increasing global demand

and shortages of commodities such as oil, and better prepared to deal with potential global food scarcities." That needs to be explained, and its limits should be appreciated.

Worldwatch defines resilience as "the ability of natural or human systems to survive in the face of great change."

> To be resilient, a system must be able to adapt to changing circumstances and develop new ways to thrive. In ecological terms, resilience has been used to describe the ability of natural systems to return to equilibrium after adapting to changes. In climate change, resilience can also convey the capacity and ability of society to make necessary adaptations to a changing world—and not necessarily structures that will carry forward the status quo. In this perspective, resilience affords an opportunity to make systemic changes during adaptation, such as addressing social inequalities. (Worldwatch 2009, 203)

On this basis, resilience in a food system would mean being able to choose from a variety of alternative food sources, and being ready to jump from one to another in an agile way with changing conditions. Resiliency is different from self-sufficiency. Food security, in the sense of ensuring access to food under all conditions, comes mainly from resiliency, not self-sufficiency.

Hawai'i should have a variety of food sources available so that if one fails or weakens, it would be possible to shift to other sources. Hawai'i has already done that on a regular basis for fresh produce, with wholesalers jumping around to different sources, opportunistically. Increasing Hawai'i's capacity to produce its own food would increase its resiliency to the extent that it added another source of food. However, if it displaced other sources, the result could be decreased resiliency. Hawai'i should not pursue food self-sufficiency to the extent that it allows its contacts with other sources of food to wither away. Just as Hawai'i should not be overly dependent on imports, it should not be overly dependent on its own production.

Disasters

Hawai'i has had a long run of good fortune, but it is not immune from disasters. Given its huge dependence on imports, the state has to be especially concerned about possible disruptions in transport to the islands. In 1949, when it was still a territory of the United States, Hawai'i suffered from a shipping strike, and wondered aloud about what the US government would do to help (*Time* 1949). Hawai'i is now a state of the United States, but it is still not clear what help the US government would offer if Hawai'i, and possibly the US government itself, encountered some sort of extreme situation.

Disaster is defined by the United Nations International Strategy for Disaster Reduction (UNISDR) as:

> A serious disruption of the functioning of a community or a society causing widespread human, material, economic or environmental losses which exceed the ability of the affected community or society to cope using its own resources. (UNISDR 2006)

Resilience can be understood as the capacity to make adaptations to the existing food system in response to changes in the physical or economic environment. In dealing with slow and permanent changes, it is about creating a new kind of "normal." In disaster planning, however, the concern is to find ways to prepare for quick changes, especially unanticipated quick changes. Usually disaster planning is based on the hope that the impact will be of short duration, and it will be possible to return to the same basic food system that existed before the disaster. In extreme disasters that system may need to be reconfigured with great urgency.

Hawai'i has not yet had major problems with its overall food supply, but there is a need for concern because Hawai'i imports so much of its food. Disruptions to that delivery system could be disastrous, especially if the disruption is sudden and Hawai'i is unprepared.

Production

Preserving and expanding Hawai'i's farm acreage alone would not be enough to ensure future food security. If there were to be a sudden cutoff in imported food, we would need a rapid switchover from production of nonfoods and nutritionally unimportant foods (e.g., coffee, macadamia nuts, and herbs) to basic foods to ensure that everyone is well nourished. Plans should be made well in advance to facilitate such a conversion if and when it should become necessary. Historical wartime mobilizations suggest the possibilities for rapidly increasing local production of basic food.

In extreme emergencies, national and local governments might not be able to cope. Thus, some people focus on household and local food production, taking measures that are independent of government initiatives. For many people this is an ideological issue, based on the premise that even in good times, families and local communities ought to depend mainly on foods that they themselves produce. Some survivalists take this to an extreme, and many others do these things in a more limited way.

Storing Food

To deal with emergencies, it is important to work not only on food production but also on food storage, at the state level, in communities, and in households. Household food storage is increasingly important because the major food sellers no longer maintain large warehouses. The "just in time" delivery system has sharply reduced the merchants' need for warehouses, so now, if shipping to the state were to be suddenly cut off, the supply of food would last no more than a few days. Many people store nonperishables and water supplies in their basements or closets. Many groceries now sell specially designed emergency food supplies to be stored at home.

Historically, many places have identified particular famine foods. Sweet potatoes are especially good for this purpose, and could be grown in many places that are otherwise unused, such as forests and meadows (Kristof 2010). In Hawai'i, 'ulu (breadfruit) played an important role in protection against disasters:

> Legend traces its origin to a time of famine when Kū, the god of building and war, buried himself in the earth near his home. He later turned into an 'ulu tree so that his wife and children would not starve. Because of this, 'ulu was considered "famine" food. 'Ulu was one of the plants Polynesians brought in their sailing canoes when they discovered the Hawaiian Islands. It is a staple food throughout the Pacific, and in ancient Hawai'i it was a crop of much greater nutritional, cultural, and political significance. (Pukui 1983)

Hawai'i should prepare for many different kinds of contingencies. The state could be deeply affected by disasters locally, and also by disasters elsewhere if they interrupt the flow of food to Hawai'i. For example, if bees stopped doing their pollination work in Hawai'i, its agriculture system could weaken or even collapse. It would then have to import more food. If bees quit working in some places outside Hawai'i, it could import from other places. If bees quit everywhere, everyone would be in trouble.

Hawai'i should be concerned not only about actual shortages, but also about anticipated shortages as well. If rumors build up about a possible shipping interruption, there could be a run on food stores. There is no evident governmental plan for dealing with hoarding before, during, or after disaster events. At the global level, speculation in food commodities can be viewed as another form of hoarding, one that could result in increased food insecurity for many people. The great global land grab, in which rich countries are gaining control over poor countries' agricultural resources to ensure their own future food security, is another form of hoarding at the global level (CHR&GJ 2010). Hawai'i is not immune from such forces. To illustrate, if Hawai'i's regular sources of rice suddenly diverted their production to other buyers, Hawai'i would be in serious trouble. The state is not likely to restore Waikiki to rice production. In prolonged emergencies there might be a need for food rationing of some sort. In extreme situations there might be a need for martial law, as there was in the 1940s (Bennett 1942).

As indicated above, the United Nations defines disasters as situations that are beyond the coping abilities of any particular place. This means that in disaster planning we must go beyond strengthening the capacities of individuals, families, and communities. There is a need to work out systems for assistance among different places. This could mean systems of support from one Ahupua'a[2] to another, one island to another, or the entire State of Hawai'i to the United States, other nations, or the global community as a whole.

Despite Hawai'i's vulnerabilities, these relations have not been worked out with the clarity and foresight that is needed. If Hawai'i had a big problem with its food supply, it might be able to get help from the outside, but there are huge uncertainties. Who would the aid come from, and on what terms? Some people might assume that the US government would come to Hawai'i's assistance under various contingencies, but we do not know for sure. How long would the US government help? In what ways? Are there commitments in writing? What if the entire United States faces an emergency and becomes unable to come to Hawai'i's assistance? Where else could Hawai'i direct its appeals for help?

Ideas on how to approach these issues are suggested by the Model Intrastate Mutual Aid Legislation, available through the Hawai'i State Civil Defense website at http://www.scd.hawaii.gov/nims.html. Much work remains to be done on this. The State Civil Defense system focuses on hazards such as tsunamis and earthquakes, but does not give attention to things such as shipping interruptions or disruptions in the state's agriculture. Attention should be given to food-related dimensions of disasters such as tsunamis and earthquakes. In all disasters there are food-related problems that begin to show up as soon as the warnings begin, with runs on the stores. Any sort of prolonged disaster would raise serious concerns about food. There are also possibilities for food-centered disasters that have nothing to do with tsunamis and earthquakes. Plans should be made for dealing with food crises regardless of the cause of the disruption. The benefits would far outweigh the costs. However, there is currently no clear mandate for any agency of the state government to undertake this work.

Poverty

Poverty-based food insecurity occurs in high-income as well as low-income countries. A great deal could be learned from the way it is addressed in other high-income countries (e.g., Sydney Food 2011). In many countries the problem of the food security of the poor is given little attention, but it occurs in some degree everywhere. In the United States the federal government has been undertaking regular studies of food insecurity, focusing on the type that is associated with poverty. Adopting the methods of the US Department of Agriculture (USDA), Hawai'i's state Department of Health in 2001 went into the issues much more deeply. It concluded: "[F]ood insecurity was prevalent in Hawai'i: one in six (16.4%) households and 1 in 5 (19.2%) individuals experienced either being at risk of hunger or experiencing hunger in 1999–2000. The poor, children, single adult households, and Pacific Islanders were particularly vulnerable."

In Waimanalo, Wai'anae, Puna, Ka'a'awa, and Moloka'i, more than 30 percent of the people lived in households that were not sure how they would get their food. Because of the high cost of living, many people who are not poor, officially, suffer from food insecurity (HDOH 2001). The official poverty rate in Hawai'i hovers around 10 percent of the total population. Among the different ethnic groups, Native Hawaiians have the lowest average family income (Kana'iaupuni et al. 2005). The impact is clear in the distribution of food insecurity in Hawai'i. Another study on poverty-related food insecurity in Hawai'i was published in 2002 (Giles, Zaman, and Derrickson 2002). It used the USDA framework, but went further by sketching out a proposed Community Food Security Plan. It emphasized the need for action by the state legislature, and described several bills that were submitted to the legislature but not passed.

According to USDA data estimates, averaging for the years 2007–2009, 11.4 percent of Hawai'i's households had low or very low food security, compared to 13.5 percent for the United States as a whole; 3.9 percent of households in Hawai'i had very low food security, compared to 5.2 percent in the United States as a whole (USDA 2010a). Thus, Hawai'i has done relatively well. Nevertheless, poverty-based food insecurity is a persistent issue in the state, and, as indicated above, the prevalence is higher among particular groups. The USDA has had to take into account the extraordinarily high prices of food in just two states. "For residents in Alaska and Hawaii, the Thrifty Food Plan costs were adjusted upward by 19 percent and 63 percent, respectively, to reflect the higher cost of the Thrifty Food Plan in those States" (USDA 2010b, 57, note 3). Higher food prices mean greater food insecurity for much of the state's population, not just for the very poor.

The Hawai'i Foodbank describes itself as

> the only nonprofit 501(c)3 agency in the state of Hawaii that collects, warehouses and distributes mass quantities of both perishable and nonperishable food to 250 member agencies as well as food banks on the Big Island, Maui and Kauai.

In one year the Hawai'i Foodbank, through its cooperating agencies, served 183,500 different people in the state, including more than 55,000 children and more than 11,000 seniors. The foodbank said:

- 79 percent of client households served are food insecure, meaning they do not always know where they will find their next meal.
- 43 percent of these client households are experiencing food insecurity with hunger, meaning they are sometimes completely without a source of food.
- 83 percent of client households with children served are also food insecure.
- Of the 183,500 people the Hawai'i Foodbank network serves:
  - 79 percent of households have incomes below the federal poverty line .
  - The average monthly income for client households is $850.
  - 42 percent of households have one or more adults who is working (Hawai'i Foodbank 2010).

Each year the foodbank organizes large-scale campaigns to collect non-perishable food products from many different donors. It then provides food at little or no cost to agencies such as Aloha Harvest, the Institute for Human Services, Salvation Army, Waikiki Health Center, River of Life Mission, Kau Kau Wagon, Harbor House, and many church pantries so that they can respond to food insecurity and related problems. The programs that hand out food to the needy do a good job of tiding people over, but many unmet needs remain. The foodbank periodically raises the alarm about widespread hunger in the state when it conducts its food collection drives, but historically the state government has said very little about the issue. This may leave people uncertain as to whether it is really a serious problem in Hawai'i.

The state government administers hundreds of millions of dollars that come into the state each year for federally funded nutrition programs such as school meals, the Supplemental Nutrition Assistance Program, SNAP (formerly Food Stamps), and the Special Supplemental Nutrition Program for Women, Infants, and Children, commonly known as WIC. However, apart from that, the state has not addressed the problem of poverty-based food insecurity. It has taken little notice of the data on food insecurity in Hawai'i that are provided each year by the US Department of Agriculture. The state Department of Health used to include food-security questions in its annual health survey, but it no longer does that. And its 2001 study on "Hunger and Food Insecurity in Hawai'i" remains not updated.

Poverty-based food insecurity in Hawai'i is not high by global standards, but it exists and contradicts the image the state tries to portray of the quality of life on the islands. Hawai'i does not provide a strong safety net for all its people. State officials might feel that the coverage by federally funded programs such as SNAP and WIC, together with the work of the nongovernmental organizations, is enough to meet the needs. However, there is a need to determine whether that is so, and to consider what should be done for those who fall through the cracks.

The state's inattention to poverty-based food security issue may be partly due to the fear that dealing with it could be costly. However, there are many helpful things that could be done at low cost. The state could do more to pursue federal grants for community nutrition, such as those available through the US Department of Agriculture. The state's modest support for the local nongovernmental groups working on the issue, such as the Hawai'i Foodbank, seems to have yielded considerable benefits for a very small investment.

Many people who are eligible for SNAP and WIC do not take advantage of their services. The state, working together with interested nongovernmental organizations, could encourage more eligible people to apply. Hawai'i's legislature could learn from the ways in which other states invest a small amount of resources to help their people take full advantage of federal programs (e.g., Illinois General Assembly 2004). There are other opportunities to draw in benefits for the poor, even if they are not specifically food-oriented. For example, it has been estimated that as many as 34,000 taxpayers in Hawai'i may not be applying for the Earned Income Tax Credit to which they are entitled. The Family and Individual Self-Sufficiency Program at the Hawai'i Alliance for Community Based Economic Development offers help along these lines, but more could be done (Tanna 2010).

When the state is going through a difficult time economically and cutting back on public services, it should give more attention to the food security issue, not less. This does not necessarily mean that the state has to provide more direct services. It should monitor the issue, and call for help where it is needed. The challenge is not to feed people, but to see to it that they live under conditions in which they can provide for themselves. Dignity comes from providing for yourself and your family, not from standing in a soup-kitchen line. All able-bodied people should have decent opportunities to take care of themselves. Regardless of whether we draw on federal resources or charitable giving or local farmers' markets, the state government should take the responsibility to ensure that no one in the state remains food insecure.

Food Policy Bodies

It is not only poverty-related food security that the state has ignored. Until recently, the state government in Hawai'i has given little attention to the security of the overall food supply, and it has not done disaster planning related to possible food crises. Whether we are concerned with sudden-onset disasters or threats to the food supply that come with slow climate change and increasing energy costs, there are compelling reasons for serious planning. Hawai'i's food system should be designed to be as resilient as possible so that it is prepared to deal with all sorts of changes in conditions. Ensuring good nutrition for all segments of the population under all conditions is a challenge that requires sustained attention.

In 2002 and early 2003, with prodding from interested citizens, the state legislature asked the Office of Planning in the state Department of Business, Economic Development, and Tourism to convene a Food Security Task Force, to examine the best ways to ensure food security for Hawaiʻi's people. As a result of that group's work, in 2003 the legislature considered bills to create a permanent state Food Security Council. As stated in the conclusion of the Task Force's report:

> Hawaiʻi has no State, county or local food policy council to coordinate or oversee food security activities. Without State policies, objectives, or goals to guide State actions, no organization can effectively coordinate assistance programs, conduct ongoing monitoring, or spearhead integrated planning programs. With an adequate State match (funds, personnel), on an on-going basis, the State could leverage available federal dollars for food security coordination, food stamp outreach and education, and farmers markets initiatives, which can then be used to enhance food security and put food dollars into the pockets of the needy, local farmers and food retailers thereby spurring our economy from the ground up. (Food Security Task Force 2003, 14)

The idea was that the council, including both government officials and private citizens, would envision a food secure Hawaiʻi, and then try to figure out how to get there. The council would bring together all concerned parties to formulate a coherent strategy for identifying and addressing the issues. However, the legislature did not approve the proposal.

Given the persistent need to strengthen Hawaiʻi's food security, interested individuals and organizations gathered together in November 2010 to establish a nongovernmental, community-based Hawaiʻi Food Policy Council (HFPC) (Lukens 2010). As explained at its website (http://www.hawaiifoodpolicycouncil.org/) and its Facebook page (http://www.facebook.com/HawaiiFPC), the HFPC's primary role is to provide a forum for exploring the major food security issues confronting the state.

Without the engagement of the state government, the HFPC's capacities would be very limited. There is a need for an interagency unit in the state government that would have primary responsibility of ensuring food security for all parts of the state's population under all conditions. This unit could work together with the community-based HFPC, and serve as a major channel through which the government would hear the concerns of the people. Hawaiʻi's government and people need to act together to strengthen the local food system, and address the full range of food security issues that confront it.

Notes

1. The Maritime Trade Act of 1920, § 27.
2. Geographic land division based on Hawaiian history and ecology; see, e.g., Chapter 3 in this volume.

References

Bennett, M. K. 1942. "Hawaii's Food Situation." *Far Eastern Survey*, Vol. 11, no. 16 (August 10): 173–175.

CHR&GJ. 2010. *Foreign Land Deals and Human Rights: Case Studies on Agricultural and Biofuel Investment.* New York: Center for Human Rights and Global Justice, New York University School of Law. http://www.business-humanrights.org/Links/Repository/1003276.

Conrow, Joan. 2009. "A Seed of Doubt." *Honolulu Weekly*, Vol. 19 (April 8–14): 6–7. http://honoluluweekly.com/cover/2009/04/a-seed-of-doubt/.

Dean, Adam. 2007. *Local Produce vs. Global Trade.* New York: Carnegie Council: Policy Innovations. October 25. http://www.policyinnovations.org/ideas/briefings/data/local_global.

Desrochers, Pierre, and Hiroko Shimizu. 2008. *Yes, We Have No Bananas: A Critique of the "Food Miles" Perspective.* Arlington, VA: Mercatus Center, George Mason University. http://mercatus.org/publication/yes-we-have-no-bananas-critique-food-miles-perspective.

DeWeert, Sarah. 2009. "Is Local Food Better?" *Worldwatch.* May/June. http://www.worldwatch.org/node/6064?emc=el&m=227941&l=4&v=67949a0ab6.

FAO. 2009. *The State of Food Insecurity in the World.* Rome: Food and Agriculture Organization of the United Nations. http://www.fao.org/docrep/012/i0876e/i0876e00.htm.

Food Security Task Force. 2003. *A Report to the Legislature on SCR 75, SD1, HD1, 2002.* Honolulu: Office of Planning, State of Hawai'i. http://hawaii.gov/dbedt/op/fstfr_2003.pdf.

Giles, Catherine, Shireen Zaman, and Joda P. Derrickson. 2002. *Hawai'i Community Food Security Needs Assessment.* Washington and Honolulu: Congressional Hunger Center and Full Plate. http://www.wellnessconsultantandresourcecenter.com/files/3553774/uploaded/HIComFoodNeedsAssess2003.pdf.

Gomes, Andrew. 2011. "State Farm Revenue Near Peak." *Honolulu Star-Advertiser.* January 16, D1. http://www.staradvertiser.com/business/20110116_State_farm_revenue_near_peak.html.

Haraguchi, Karol, and Linda Menton. 1987. *Rice in Hawaii: A Guide to Historical Resources.* Honolulu: State Foundation on Culture and the Arts and Hawaiian Historical Society.

Hawai'i Foodbank. 2010. *Hunger Facts.* Honolulu: Hawai'i Foodbank. http://www.hawaiifoodbank.org/page4.aspx.

Hawai'i Food Manufacturers Association. 2011. http://www.foodsofhawaii.com/.

Hawai'i State Legislature. 2012. Hawai'i State Legislature: Bill Status and Documents. http://www.capitol.hawaii.gov/.

HDOH. 2001. *Hunger and Food Insecurity in Hawai'i: Baseline Estimates.* Honolulu: Hawai'i Department of Health. http://www.hawaii.gov/health/statistics/hhs/index.html/pdf/specfood.pdf.

Illinois General Assembly. 2004. *Public Aid: (305 ILCS 42/) Nutrition Outreach and Public Education Act.* http://www.ilga.gov/legislation/ilcs/ilcs3.asp?ActID=2476&ChapterID=28.

Kana'iaupuni, Shawn Malia, Nolan J. Malone, and Koren Ishibashi. 2005. *Income and Poverty among Native Hawaiians: Summary of Ka Huaka'i Findings.* Honolulu: Kamehameha Schools. http://www.ksbe.edu/spi/PDFS/Reports/Demography_Well-being/05_06_5.pdf.

Kristof, Nicholas D. 2010. "Bless the Orange Sweet Potato." *New York Times.* November 24. http://www.nytimes.com/2010/11/25/opinion/25kristof.html?_r=1&emc=tnt&tntemail1=y.

Lee, C. N., and H. C. "Skip" Bittenbender. 2007. "Agriculture." In *Hawai'i 2050: Building a Shared Future: Issue Book.* http://www.hipaonline.com/pdf/HI2050_Issue_Book.pdf.

Leung, PingSun, and Matthew Loke. 2008. *Economic Impacts of Increasing Hawaii's Food Self-Sufficiency.* Honolulu: Cooperative Extension Service, College of Tropical Agriculture and Human Resources, University of Hawai'i at Mānoa.

Lukens, Ashley. 2010. "Democratizing Food: Hawai'i's Future Food Systems." *Honolulu Weekly.* November 24. http://honoluluweekly.com/feature/2010/11/democratizing-food/.

McWilliams, James E. 2007. "Food That Travels Well." *New York Times.* August 6. http://www.nytimes.com/2007/08/06/opinion/06mcwilliams.html?_r=1&oref=slogin.

Nakaso, Dan. 2011. "Basil Bane Putting Bit on Business." *Honolulu Star-Advertiser.* February 3. http://www.staradvertiser.com/news/20110203_Basil_bane_putting_bite_on_business.html.

Page, Christina, Bony Lionel, and Laura Schewel. 2007. *Island of Hawai'i Whole System Project: Phase 1 Report.* Rocky Mountain Institute. http://www.kohalacenter.org/pdf/hi_wsp_2.pdf.

Pukui, Mary Kawena. 1983. *'Ōlelo No'eau: Hawaiian Proverbs and Poetical Sayings.* Honolulu: Bishop Museum Press.

Roberts, Paul. 2009. "Spoiled." *Mother Jones,* Vol. 24, no. 2 (March/April): 28–36.

Schmitt, Robert C. 1970. "Famine Mortality in Hawai'i," *Journal of Pacific History,* Vol. 5.

Singer, Peter, and Jim Mason. 2007. *The Way We Eat: The Ethics of What We Eat and How to Make Better Choices.* Emmaus, PA: Rodale Press.

Southichack, Mana. 2007. *Inshipment Trend and Its Implications on Hawaii's Food Security.* Honolulu: Hawaii Department of Agriculture. http://www.kohalacenter.org/pdf/HDOA_hawaii_food_security.pdf.

Sydney Food. 2011. *Understanding Food Insecurity: Why Families Go Hungry in an Affluent Society.* Sydney, Australia: Sydney Food Fairness Alliance and Food Fairness

Illawarra. http://www.mbcommunitygardens.com.au/main/images/stories/pdf/understanding-food-insecurity.pdf.

Tanna, Wayne M. 2010. "Earning Income Credit Is Bright Start of Tax Policy." *Honolulu Star- Advertiser.* January 27. http://www.staradvertiser.com/editorials/20110127_earned_income_credit_is_bright_star_of_tax_policy.html.

UNISDR. 2006. *Terminology: Basic Terms of Disaster Risk Reduction.* Geneva, Switzerland: ISDR. http://www.unisdr.org/eng/library/lib-terminology-eng-p.htm.

USDA. 2010a. *Table: Household Food Security Status by State.* Washington, DC: US Department of Agriculture. http://www.ers.usda.gov/publications/err108/statetable.htm.

———. 2010b. *Household Food Security in the United States, 2009.* Washington, DC: US Department of Agriculture. http://www.ers.usda.gov/features/householdfoodsecurity/.

———. 2011. *State Fact Sheets: Hawaii.* Washington, DC: US Department of Agriculture. http://www.ers.usda.gov/stateFacts/HI.htm.

"Who Gives a Damn?" 1949. *Time Magazine.* July 4. http://www.time.com/time/magazine/article/0,9171,888526.00.html.

Worldwatch. 2009. "Climate Change Reference Guide and Glossary." In *State of the World 2009: Into a Warming World.* Washington, DC: Worldwatch Institute.

Yonan, Alan, Jr. 2011. "Manufacturing Slump Slows." *Honolulu Star-Advertiser.* February 22, B5. http://www.staradvertiser.com/business/businessnews/20110222_Manufacturing_slump_slows.html.

3 | Kaulana O'ahu me he 'Āina Momona

Lilikalā K. Kame'eleihiwa

> Kaulana O'ahu me he 'Āina Momona Ma muli o nā Ha'awina 'Aumākua, No laila hiki ke kahea aku nei, "Hui, e hele mai 'ai, a 'ai i ka mea loa'a!"
>
> Famous is O'ahu as a land fat with food because of ancestral teachings. Therefore we can call out, "Hello, come and eat, and eat what there is!"

Alternative agro-food movement's criticism of the contemporary food system tends to include little historical reflections. Its regular starting points—criticisms of fast food, industrialization of agriculture, and the rise of obesity in the United States, for instance—take the early twentieth century as a reference point, which is seen as a better past in comparison with the troubled food system today. Its limited historical purview is unfortunate, as the foundation of the modern industrial food system was at the demise of the much earlier indigenous food systems. The colonial dispossession of indigenous peoples from their access to land, water, and human resources, critical for producing their own food, was at the root of the subsequent expansion of the industrial food systems. In this chapter, I will narrate such a story of Native Hawaiian experiences and the demise of its Ahupua'a (valley land divisions) system under colonial influence.

Specifically, I will summarize my research on the ancestral Ahupua'a system (for surface water management), which was first recorded in the 1848 Māhele Book. Subsequently, maps showing the Ahupua'a boundaries were drawn between 1855 and 1885. The making of Ahupua'a records, and later maps, was prompted by the Māhele, the legal mechanism for the privatization of land ownership that was made Hawaiian Kingdom law in 1846 (discussed below). The 1848 Māhele Book records were lists created by 252 Konohiki (land and water managers), who sat down for the first time in history to write the names of all of the lands that they managed on paper, which had previously only passed on orally. Examination of the subsequent maps of these land records shows the intricate system of land and water management that Native Hawaiians had crafted over time and how colonial forces led to its decimation that we see today.

I will argue that any attempts to move towards food democracy need to consider the impact of colonialism on indigenous food systems, and should learn from the indigenous people's ancestral and efficient management practices of water, land, and food production on small islands like O'ahu. The chapter ends with some examples of food activism that are in line with the ancestral wisdom of Native Hawaiians.

Kaulana O'ahu me he 'Āina Momona ma muli o nā Ha'awina 'Aumākua

In 1782, when Kahekilinui'ahumanu, the Mō'ī (King) of Māui, planned his invasion of O'ahu, he said O'ahu was an 'Āina Momona, or an island "fat with food" (Kamakau et al. 2002). What did he mean by this odd phrase?

He was referring to the great abundance of fresh water on O'ahu, which was used to build 113 fishponds, comprising 4,200 acres, more than the 108 fishponds found on all the other seven islands of the Hawaiian archipelago. O'ahu also had the largest fishpond, named Kuapā, also known as Keahupua fishpond, comprising 523 acres located in Maunalua, an 'ili (smaller land division) of the Ahupua'a of Waimānalo in the Moku district Ko'olaupoko (Cobb 1904). Until it was made into the Hawai'i Kai marina in the 1960s, Kuapā may have been the largest fishpond in the Pacific. Moreover, as a working fishpond in the 1800s was known to produce 300–500 pounds of fish per acre per year, O'ahu was producing 1.3 million pounds of fish per year at the minimum, not counting the fish obtained from the reefs and beyond (Keala et al. 2007). That is what made O'ahu an 'Āina Momona.

In the 1880s, 79 percent of the Ahupua'a on O'ahu were so well watered that Lo'i Kalo (wetland taro terraces) were constructed from the back of the valleys down to the shoreline. These Lo'i Kalo systems were built in sixty-three out of the eighty Ahupua'a on O'ahu. Extensive wetland taro terracing was then an indicator of a large population to feed since wetland taro cultivation produces 10–15 times more food than dryland taro (Kelly 1989).

Imagine all of the Mānoa valley was filled with wetland taro from the foot of Mānoa Falls all the way down to Kalākaua Avenue on Waikīkī beach. Nearly every valley on O'ahu was like that. In 1792, Captain George Vancouver landed on Waikīkī, seeking fresh water to replenish his ship supplies. He wrote in his journal about the Waikīkī taro fields:

> Our guides led us to the northward through the village [Waikīkī], to an exceedingly well-made causeway, about twelve feet broad, with a ditch on each side. This opened to our view a spacious plain . . . the major part appeared divided into fields of irregular shape and figure, which were separated from each other by low stone walls, and were in a very high state of cultivation. These several portions of land were planted with the eddo, or taro root, in different stages of inundation, none being perfectly dry, and some from three to six or seven inches under water. The causeway led us near a mile from the beach, at the end of which was the water we were in quest of. In this excursion we found the land in a high state of cultivation, mostly under immediate crops of taro and abounding with a variety of water fowl, chiefly of the duck kind . . .
>
> At Woahoo [O'ahu], nature seems only to have acted a common part in her dispensations of vegetable food for the service of man; and to have almost confined them to the taro plant, the raising of which is attested with much care, ingenuity, and manual labour. (Vancouver 1798, 163–164)

Accompanying Captain Vancouver was the British naturalist Archibald Menzies, who wrote in his journal:

> We pursued a pleasing path back into the plantations which was nearly level and very extensive, and laid out with great neatness into little fields planted with taro, yams, sweet potatoes and the cloth plant. These in many cases, were divided by little banks on which grew the sugar cane and a species of Draccena without the aid of much cultivation, and the whole watered in a most ingenious manner by dividing the general stream into little aqueducts leading in various directions so as to supply the most distant fields at pleasure, and the soil seems to repay the labour and industry of these people by the luxuriancy of its productions. (Menzies 1920, 23–24)

Some twenty-three years later, in 1815, not much had changed in Waikīkī, as the Russian Captain Otto Von Kotzebue observed:

> The luxuriant taro-fields, which might be properly called taro-lake, attracted my attention. Each of these consisted of about one hundred and sixty square feet, forms a regular square, and walled round with stones, like our basins. This field or tank contained two feet of water, in whose slimy bottom the taro was planted, as it only grows in moist places. Each had two sluices. One to receive, and the other to let out, the water into the next field, whence it was carried farther. The fields became gradually lower, and the same water, which was taken from a high spring or brook, was capable of watering a whole plantation. When the taro is planted, the water is lowered to half a foot, and the slip of a gathered plant stuck into the slime, where it immediately takes root, and is reaped after three months. The taro requires much room, having strong roots; it strikes forth long stalks and great leaves, which appear to swim on the water. In the spaces between the fields, which were between three and six feet broad, arc pleasant shady walks, planted on both sides with sugar-cane or bananas.
>
> They also use the taro-fields as fishponds. In the same manner as they keep the river-fish here, they keep the fish in the sea, where they sometimes use the outer coral-reefs, and form from them to the shore a wall of coral stones, thus making fish-preserves in the sea. Such a preserve requires much labour, but by no means so much art as the taro-fields, which serve for both purposes. I have seen whole mountains covered with these fields, through which the water flowed gradually down, each sluice forming a cascade, and falling between sugar-canes and banana-trees into the next tank. Sugar plantations, taro-fields, and far-scattered plantations succeeded each other on our road. . . . (Kotzebue 1821, 102)

Sixty-six years later in 1881, Waikīkī, which means spouting water in reference to its many subterranean springs, was still a marvelous network of waterways filling Lo'i Kalo and fishponds with an astonishing volume of fresh water, which was used to intensify food production to feed the population (figure 3.1).

What happened to Waikīkī? What happened to O'ahu? What happened to the 'Āina Momona? The abundance of food depicted in the historical records is hard to imagine from what we see in today's Hawai'i, which is dependent upon imported food from outside the islands and has lost much of its indigenous food production.

Since our Hawaiian ancestors have lived on the small islands of the Hawaiian archipelago for the past one hundred generations, or roughly two thousand years, with such great food producing ingenuity, we might want to consider how they did it. Moreover, Hawai'i had an estimated population of about 1 million at the point of contact in 1778 (Stannard

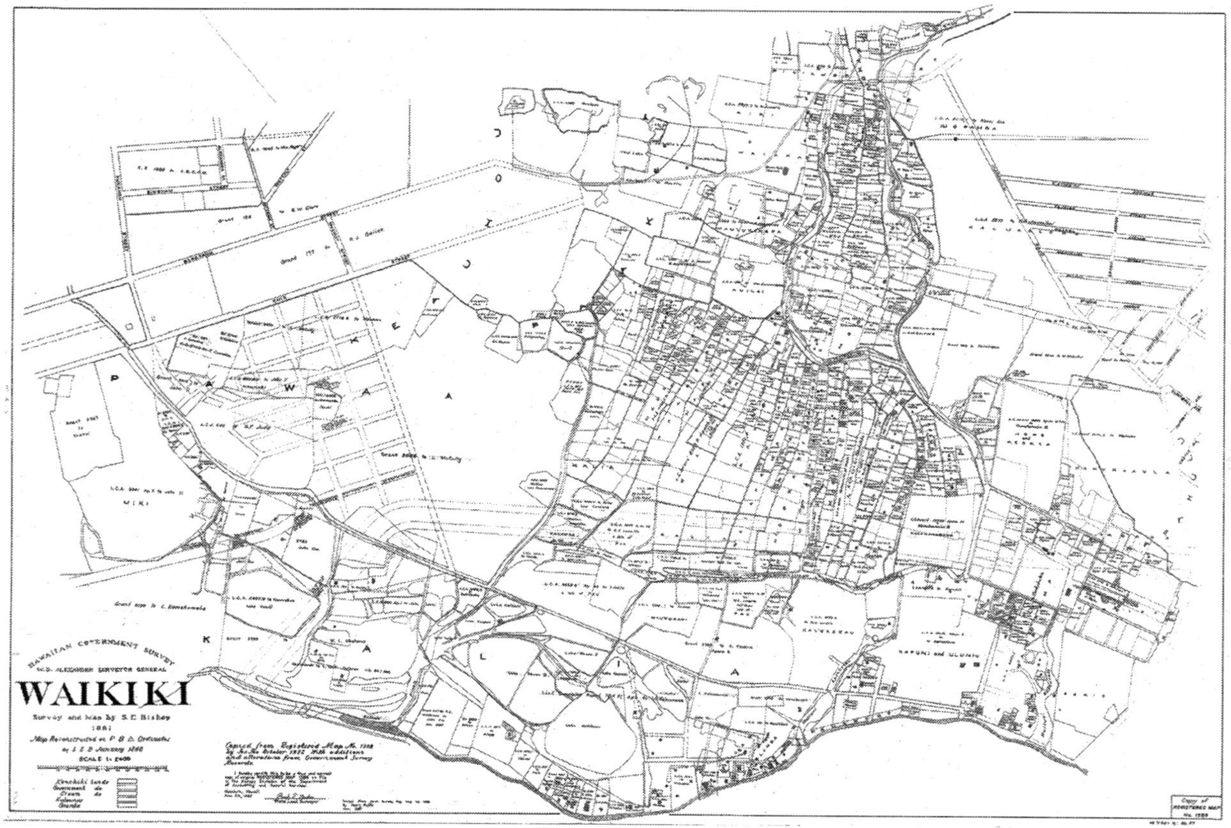

FIGURE 3.1. Map of Waikīkī in 1881 by Sereno Bishop, with the dark lines showing inland fishponds and ʻauwai (water channels), flowing from the mountain to the sea, that fed the numerous wetland taro fields of the area. Register Map No. 1398, Department of Land and Natural Resources.

1990; Kame'eleihiwa 1992), which roughly mirrors the current 1.3 million population of the State of Hawai'i. How did the ancestors feed everyone when they could not import 90 percent of their food from the North American continent? Let us take a look at the ancestral methods.

Ancestral Wisdom: Moku and Ahupua'a Management Systems

Our Native Hawaiian ancestors divided the islands into large land districts called Moku. Each Moku district was divided further into Ahupua'a, or land areas that went from the mountain all the way down to the reefs and into the sea (Malo 1959). The ridgelines of various valleys were usually the horizontal boundaries. On O'ahu it is easy to see that the Ahupua'a boundaries actually demarcated surface water management areas. When one studies fishponds and how they were connected by underground waterways, it seems likely that the larger Moku districts were subterranean water management areas.

Kāne, the god of sun and the new knowledge of the day, was also the god of fresh water in his female form, and one of his body forms was the kalo (taro) plant. In Hawaiian mythology, Kalo is the elder brother of the Hawaiian people, as well as our favorite food—poi (pounded taro). Thus the management of water, a divine element, was also a religious calling, and prayers to the female Kāneikawaiola, Kāne of the life-giving waters as she was called, reminded us of the sanctity of water, and guided us in our very efficient use of water for the growing of food (Sproat 2009).

As a result of population increase, Ahupua'a boundaries were carefully established within the Moku districts, so that people would know they were to only use the water from their own Ahupua'a for agriculture (Nakuina 1894). Ahupua'a law taught that they were to only eat from within their own Ahupua'a, and not to go next door to another Ahupua'a for food. This practice ensured that people took very good care of their own Ahupua'a where they lived, and it ameliorated any possible disputes over water, land, and resources.[1]

Ahupua'a were further divided into smaller land divisions called 'ili, mo'o, lo'i, kuana, and mala subdivisions (Malo 1959). Each of these smaller land divisions had specific names to make it easier to track who was supervising the planting of each to avoid confusion in reference. Those different names also implied different planting systems that were employed. For instance, mala were dryland gardens that did not need to be irrigated, whereas Lo'i Kalo were wetland taro terraces that resembled the more widely known rice paddies, full of flowing water, connected to rivers by a system of 'auwai (water channels) that took water into the lo'i for circulation amid the kalo plants, and then delivered it back into the river. In this way folks downstream could get as much water as they needed (Nakuina 1894).

Over the one hundred generations that Native Hawaiian ancestors lived in this archipelago, they became excellent water managers, especially for the production of food, and their favorite food was poi made from kalo. Not only is kalo extremely nutritious, it is also rich in calcium, preventing tooth decay and creating strong bones. Other vegetable crops that grew easily in Hawai'i were 'ulu (breadfruit), mai'a (banana), and 'uala (sweet potato). 'Uala is actually the most wondrous of those crops because unlike kalo, which takes eighteen months on average to mature, 'uala is a 3–6-month crop that can produce hundreds of pounds in a small area, like a mo'o land division. Although 'uala originates from South America, where it is called kumara (cognate to 'uala), it seems, by ancestral accounts, to have come into the Central Polynesian island of Ra'iātea about two thousand years ago in association with the god Lono (Henry 1928). From there it spread to all of Polynesia along with the worship of Lono. Like the wetland farming of kalo, 'uala growing on slopes with good drainage can feed a large population (Handy et al. 1995).

The larger the Ahupua'a, the more the 'ili subdivisions by which it was divided, with Konohiki (water managers) overseeing every level of subdivision. In this way, every arable square foot of land was managed for the most efficient food production. Konohiki were expert water managers who were trained to understand just how the water had to flow from the mountain to the sea, so that everyone in the Ahupua'a got their fair share of water for the growing of kalo. Water was only ever diverted from the stream for the growing of kalo, and was closely regulated so that on some days some 'auwai, or water channels were closed in order for the kalo growers downstream to have a little more water on those days. On the next day the water would be restored. If anyone took water without the permission of the Konohiki, the punishment was death (Nakuina 1894).

Yearly inspections of the Ahupua'a agricultural systems began with the celebration of Makahiki (the New Year), which was presided over by the god Lono, a younger brother of the great god Kāne. Both Lono and Kāne were associated with the Haumea Earth Mother lineage of O'ahu. She is of the genealogy called Palikū (the erect cliffs), which is also an intellectual tradition of worshipping the land as the most precious female ancestor, and of exalting the female gods. The Lono priests were called Palikū priests, and they were in charge of demarcating the ridgeline boundaries of the earth that made up the Ahupua'a boundaries in order to take proper care of the Haumea Earth Mother.

As a god, Lono represented the atmospheric elements, and thus was the deity of wind, rain, fertility, and agriculture. His advent was heralded by the rising of the Makali'i (Pleiades) Constellation on the eastern horizon after dark (Malo 1959). While the Makahiki new year ceremonies, and the subsequent coming of Lono, were celebrated all over Polynesia with the rise of Makali'i stars, different practices arose in Hawai'i (Kame'eleihiwa 2009). Although the god Lono was worshipped in all other parts of Polynesia, only in the Hawaiian archipelago did the Lono priests, accompanied by the god Lono, make a clockwise circuit of the island, stopping at each Ahupua'a boundary to receive food tribute, which included an inspection of the food production efficiency of the Ahupua'a (Malo 1959). If the Ahupua'a was deemed to be underproducing, then another Konohiki who knew how to be more efficient in water management would replace the head Konohiki. This kind of Makahiki ceremony was called Lonoikamakahiki, after the Lonomākua (Lono-parent) traditions that first arose at Kualoa, O'ahu, land named for the Akualoa (Long God) wooden depiction of Lono. The clockwise Lonoikamakahiki ceremonies began on O'ahu, as did the Ahupua'a water management system, in order to make annual inspections of wetland food production a religious event (Kame'eleihiwa 2009).

The Ahupua'a pig altar, constructed on each Ahupua'a boundary along the main road that encircled the island, represented the god Lono, as the pig is a body form of Lono. Thus Lono was in attendance at all times, and even between yearly Makahiki celebrations. Moreover, Kamapua'a, literally the pig child, is the 'Aumākua (family guardian) of Lo'i Kalo farmers, who would put the navel of their new-born sons in the carved pig head on the pig altar to ensure that the next generation would be excellent kalo farmers, instead of the 'Aumākua of dryland kalo planters.[2]

It was the kuleana (responsibility) of the Mō'ī (King or Queen of the island) and the Ali'i Nui (high chiefs), both male and female, as the administrators of the land, to make sure that no one starved on the lands under their jurisdiction. Their judicious appointment of Konohiki water managers, usually their lesser lineage relatives, trained in efficient water management for maximum food production made sure that every plot of arable land was accounted for. This two-tier organizational system of Ali'i Nui and Konohiki was the secret to ensuring maximum food production (Laimana 2010). Good administration of water ensured the abundance of food, and people flocked to Ali'i Nui who excelled in this profession.

Since the Ahupua'a system, as well as the Hawaiian version of Makahiki, or Lonoikamakahiki, began on O'ahu, we shall take a closer look at how that worked on O'ahu. It was the O'ahu Mō'ī, Mā'ilikūkahi who ruled around AD 1480, that first instituted the Ahupua'a system within the Moku districts. Moku districts had been established at least nine generations earlier in the time of Mā'ilikūkahi's ancestor, Māweke, around AD 1300 (Fornander 1996; Kamakau 1996).

On O'ahu there were six Moku districts, which was typical for the larger islands of Kaua'i, and Hawai'i as well (figure 3.2).[3] On O'ahu, these six Moku districts were divided into eighty Ahupua'a. Each of the six Moku had an Ali'i Nui who served as the Konohiki of that Moku. Under him or her were Konohiki who were appointed to oversee each Ahupua'a. More often than not these positions and the training in water management were hereditary (Kame'eleihiwa 1992). Under each Ahupua'a's lead Konohiki were a team of Konohiki of 'ili divisions who worked with a set of Konohiki in charge of the Lo'i Kalo and mala within the 'ili.

On O'ahu, the Konohiki system was able to run a continuous series of Lo'i Kalo networks in sixty-three of the eighty Ahupua'a. The remaining seventeen Ahupua'a were in mala, or dryland gardens. All of these required the Konohiki to master cajoling the maka'āinana (common people), who actually farmed the land, to cooperate for the highest crop yields, to be checked by the Moku Konohiki every year during the Makahiki celebrations for the fertility god Lono. Maintaining Ahupua'a-wide Lo'i Kalo systems took a great deal of community labor on a weekly basis. On the other hand, food was plentiful and working hours were short when all worked together in the laulima (many hands) system. In 1824, the Calvinist missionary Charles Stewart, who was stationed on O'ahu, stated: "The Hawaiians are the most industrious of Polynesians; they work about 4–5 hours a day" (Stewart 1828; Handy et al. 1995).

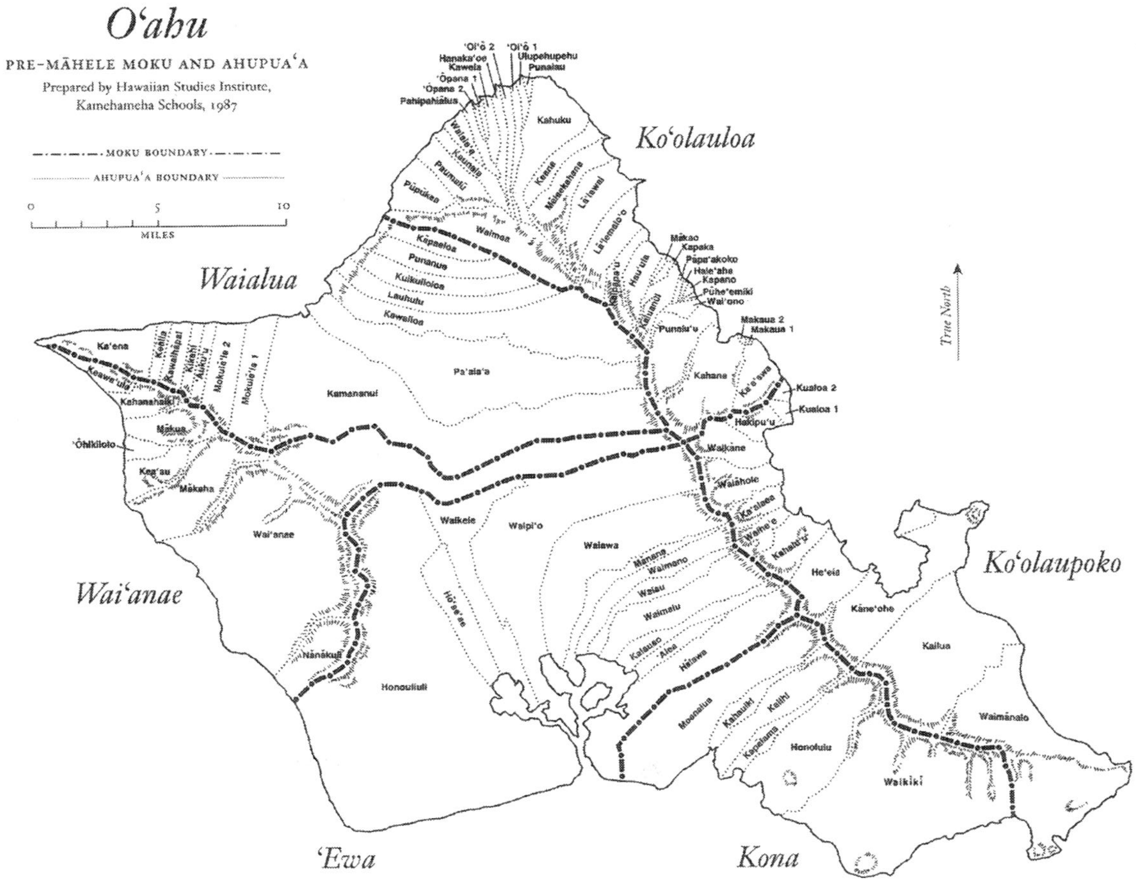

FIGURE 3.2. Map of six Moku on O'ahu and eighty Ahupua'a. Kamehameha Schools Hawaiian Studies Institute.

ʻĀina Momona: Land Fat with Food, as Defined by Fishponds

The greatest testament to the Konohiki system was the management of hundreds of acres of fishponds on Oʻahu. Fishponds are one of the most efficient ways of creating animal protein (Kelly 1989). And the knowledge of how to run fishponds was embodied in the Moʻowahine (Lizard Women) clans of Haumea, the Earth Mother goddess. Said to be born on Oʻahu on the cliffs of Nuʻuanu, Haumea and the Moʻowahine were worshiped in the Hale o Papa (female temple) where only women worshipped. According to Kamakau (1976), the first Hale o Papa was built on the cliffs of Nuʻumealani where Haumea was born. Every fishpond had a Moʻowahine guardian, and it was said that when she was absent, the fish left with her. This kind of mythology suggests that the knowledge of surface and underground water was closely guarded by Hawaiian women of the Moʻo clans who worshipped Haumea Earth Mother and other famous Moʻowahine in charge of water management. Since men could not go to the Hale o Papa female temple to worship, and learn from the goddesses, mothers would teach their sons, as well as their daughters, the intricacies of this occupation. Female temples and the exaltation of the female goddesses were only found in Hawaiʻi and not anywhere else in Polynesia, perhaps because of the great abundance of fresh water in Hawaiʻi, and especially on Oʻahu.

In 1885, there were 113 fishponds on Oʻahu out of a total of 221 fishponds in the Hawaiian archipelago. On Oʻahu, the fishponds were fed by fresh water springs, not by rivers, and some kūpuna (elders) believe that there were at least two springs per acre of fishpond, and probably more. For Oʻahu, that means there were at least 8,400 fresh water springs (Thompson 2013). Others have suggested that the fishponds were not made of a series of individual springs, but rather the entire pond was a fresh water lens, or one giant spring with the water seeping everywhere from below the surface, and only noticeable when one disturbed the fishpond floor (Naehu 2013).

Fishpond walls were built to keep out too much salt water because fish seem to grow faster in fresh water; the walls also served to protect the reefs, as coral does not grow well in the presence of fresh water. In a typical 100-acre fishpond, there would only be three makahā (sluice gates), which were occasionally opened to allow the exchange of salt water and fish with the surrounding ocean outside (Apple and Kikuchi 1975; Thompson 2013).

Studies of O'ahu fishponds suggest that Moku districts were most likely subterranean water management districts. Fishponds were connected underground from several miles away. The ancestors used to swim in those underground waterways (Thompson 2013), and in some instances such as the Pohukaina aquifer, they would paddle canoes underground from one entrance to another (Kamakau 1964).

For instance, the Kuapā pond in Maunalua (now called Hawai'i Kai), the largest fishpond at 523 acres in the Pacific, was connected underground to the Kailua fishponds of Kawainui and Ka'elepulu, which enabled the fish to migrate underground from one pond to the next (Sterling and Summers 1978). That is why Maunalua was considered a part of the Moku of Ko'olaupoko instead of the Moku of Kona, even though it faces south like the rest of Kona. Similarly, on the 1848 maps, Wai'anae Moku has an upland area called Wai'anae Uka (now called Wahiawā) that connected to the Ko'olau mountain range. As our foremost teacher of Hawaiian Ancestral Knowledge Dr. Pualani Kanahele teaches us, since mountains are the source of fresh water, the ancestors must have known that there was an underground water passage from Ko'olau to Wai'anae (see figure 3.2).

While there were 113 fishponds on O'ahu, they were not evenly located in each Ahupua'a; many Ahupua'a had several fishponds and some had none. The greatest concentrations of multiple fishponds were in the Moku districts of Ko'olaupoko, Kona, and 'Ewa. Ko'olaupoko had twelve Ahupua'a with twenty-four fishponds comprising 1,904 acres. The largest of these (as shown in figure 3.3 and figure 3.4) were Nu'upia fishpond (215 acres) in the Ahupua'a of Kāne'ohe, Kawainui fishpond (450 acres) and Ka'elepulu fishpond (280 acres) in Kailua Ahupua'a, and Kuapā fishpond (523 acres) in the Ahupua'a of Waimānalo (Sterling and Summers 1978).

As stated previously, as of 1885, within the eighty Ahupua'a there were 113 fishponds comprising a total of 4,200 acres. As a working fishpond can produce 300–500 pounds of fish per acre per year, O'ahu was raising a minimum of 1.3 million pounds of fish per year. Today, there are only eleven fishponds left on O'ahu, one in the Moku of Waialua, one in the Moku of Ko'olauloa, and nine in the Moku of Ko'olaupoko, comprising about 740 acres of fishponds that could be used for fish production; none of these is being used in a traditional manner. Most fishponds were filled in, and the

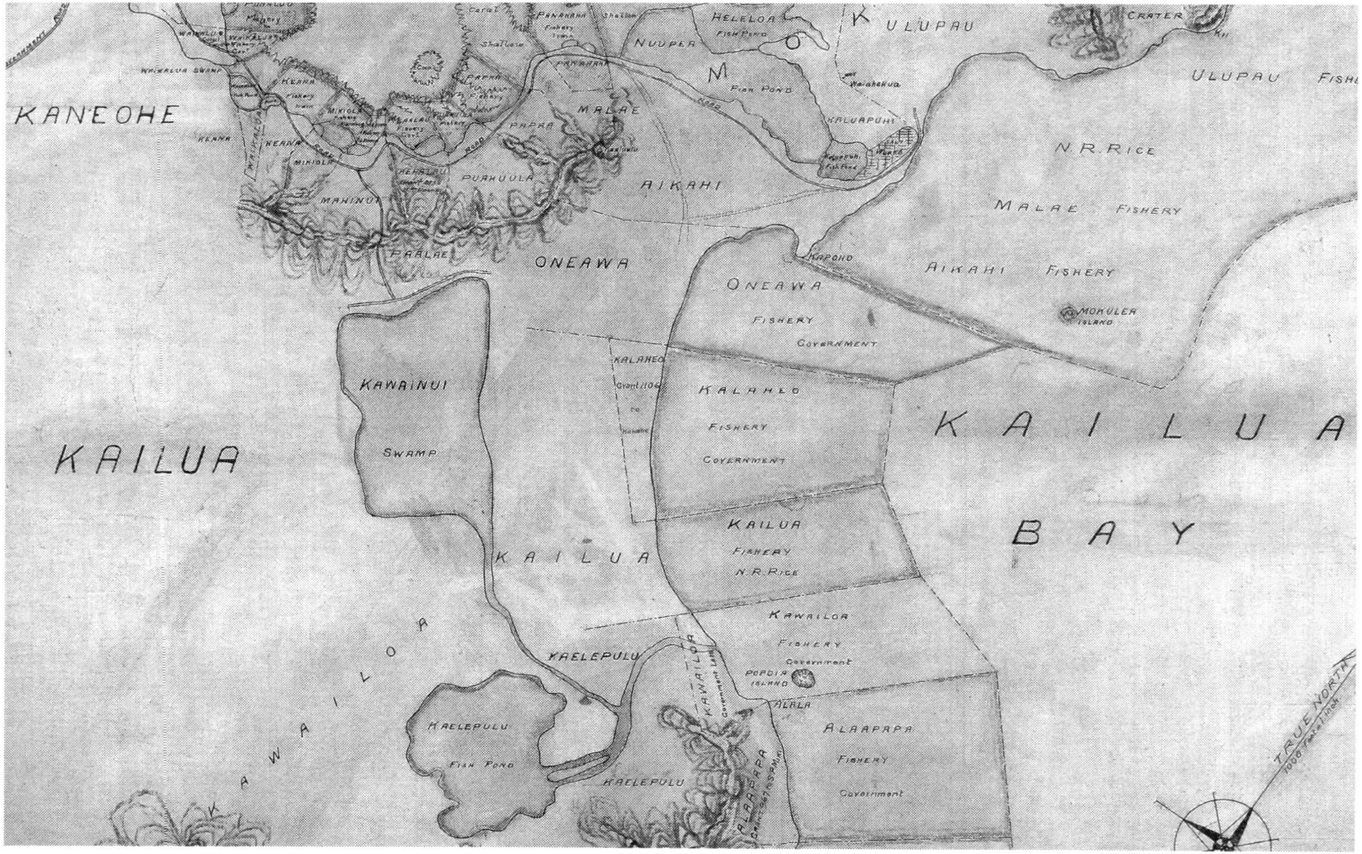

FIGURE 3.3. Oʻahu fisheries map showing the fishponds of Nuʻupia in Kāne ʻohe, and Kawainui, and Kaʻelepulu in Kailua Ahupuaʻa. Created by Monsaratt in 1913, Department of Land and Natural Resources.

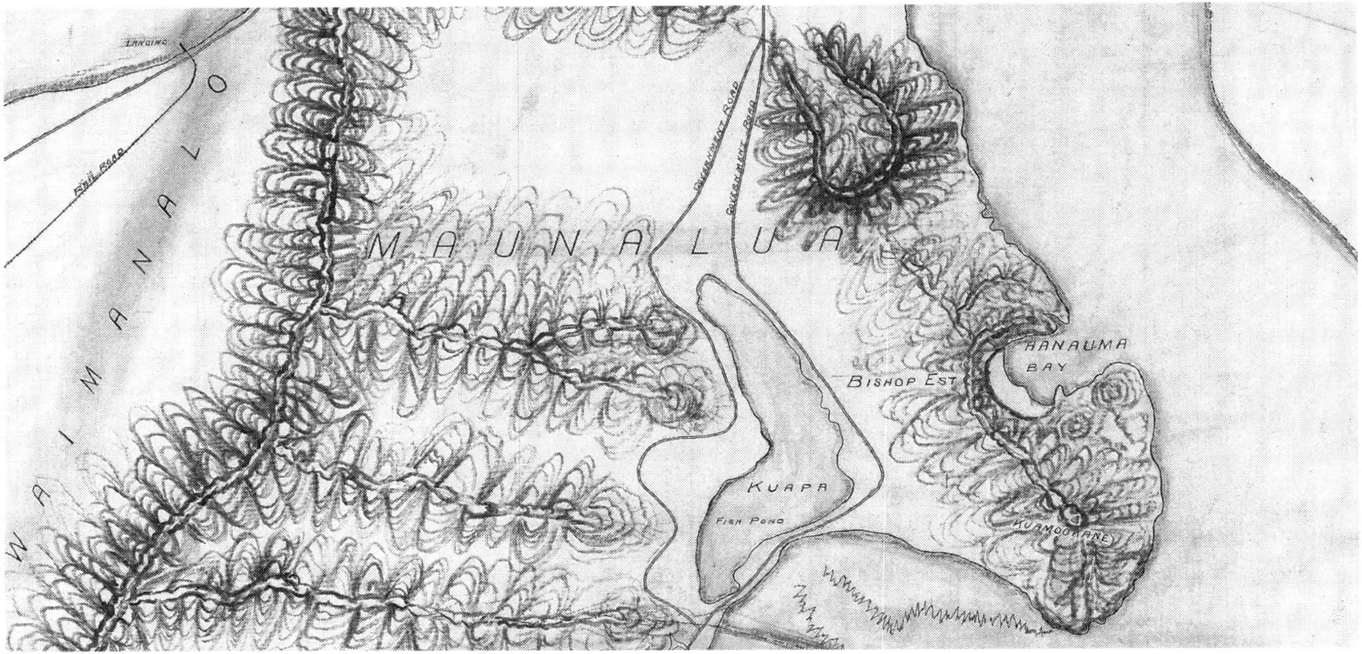

FIGURE 3.4. Oʻahu fisheries map of the 523-acre Kuapā fishpond in the ʻili of Maunalua of the Ahupuaʻa of Waimānalo. Created by Monsaratt in 1913, Department of Land and Natural Resources.

remaining ones tend to be polluted and are no longer used to produce fish. For example, there used to be forty-seven fishponds comprising 1,074 acres in the six Ahupua'a of the Kona Moku, including the extremely large Ahupua'a of Waikīkī with its eight large valleys, with the largest ponds being Lelepaua (332 acres) and Ka'ihikapu (258 acres) in the Ahupua'a of Moanalua (Cobb 1904). These two fishponds were filled in to make the reef runway at the edge of the Honolulu International Airport today.

Similarly, the 'Ewa Moku had fifteen Ahupua'a (figure 3.5), with thirty-one fishponds comprising 1,236 acres; the largest of these were Pa'au'au (320 acres) and Kuhialoko (133 acres) fishponds in Wai'awa Ahupua'a, and Hanaloa (196 acres) and E'o (137 acres) fishponds in Waipi'o Ahupua'a (Cobb 1904). However, today the only remaining fishpond Kuhialoko has become one of the primary toxic dumping areas for the American military, and can no longer be used for raising fish.

The recharge of O'ahu aquifers, so depleted by the rapid urbanization following statehood, is critical to the fresh water springs needed for fishponds. Besides the serious issue of providing fresh drinking water on O'ahu, we need to understand how to manage the surface waters of the eighty O'ahu Ahupua'a in order to replenish the aquifers, and how to manage the underground waters of the six Moku districts. Without a good understanding of the ancestral water management practices, it will be impossible to restore the fishponds to their former production capacity of 300–500 pounds of fish per acre per year. Moreover, as the long time Hawaiian environmental activist from Moloka'i Walter Ritte argues, since the fishpond mirrors the status of the Ahupua'a, it is necessary to clean up the pesticides used upland of the fishponds before we can eat the fish from them (Ritte 2013).

1848 Māhele: Effects of Capitalism and Private Ownership of Land on Native Hawaiian Food Production Efficiency

In traditional Hawai'i, there was no money, no buying and selling of land or even trade of goods. When the concept of barter was first introduced to Native Hawaiians, they considered it an extremely uncivilized behavior (Kamakau 1976). In traditional Hawaiian customs, food was freely given with aloha (love and compassion). People who planted lots of food to give

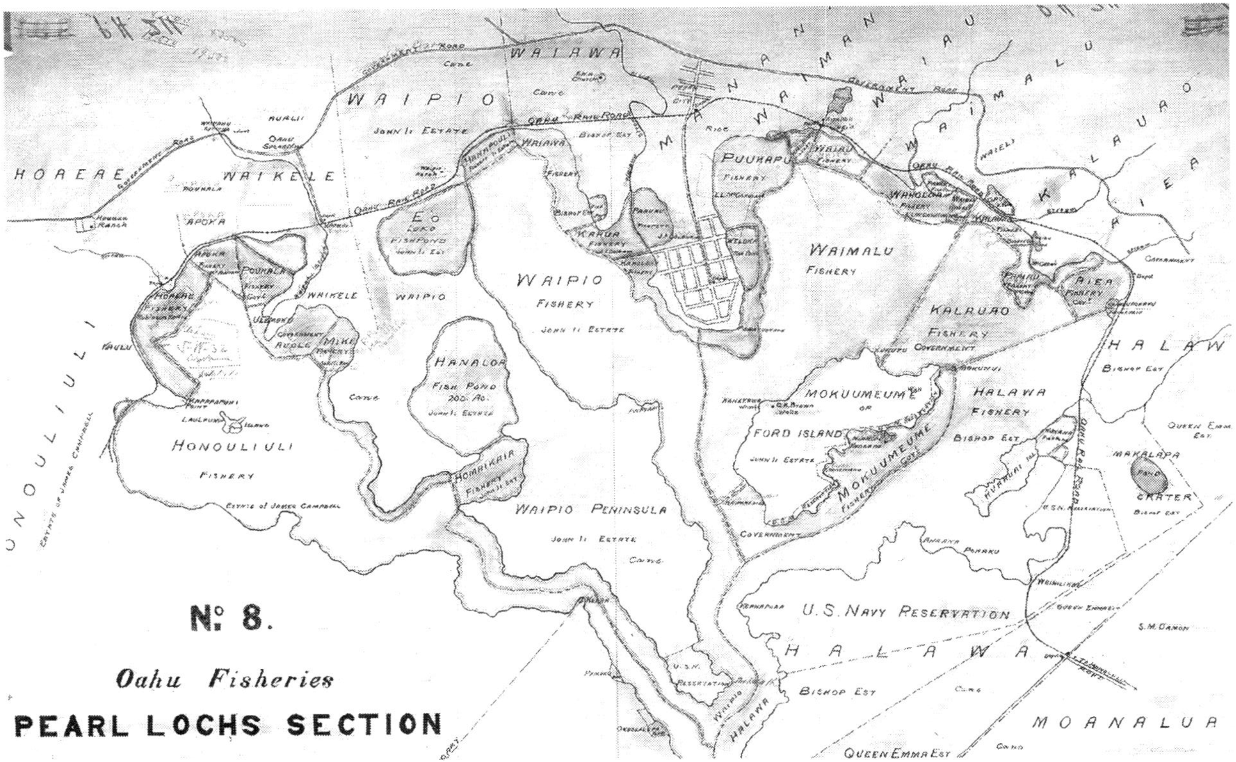

FIGURE 3.5. O'ahu fisheries map of 'Ewa Moku with the fishponds of Hanaloa, E'o, Pa'auau, and Kuhialoko. Created by Monsaratt in 1913 for the Territory of Hawai'i, Collection of the Department of Land and Natural Resources.

away were greatly admired, and they rejoiced in doing so. Giving food away and feeding strangers were considered as behaviors that would bring good luck. The idea was that the more food one gave away, the more other food would be given back. This system worked very well when there were only Hawaiians living here and when it was easy to produce lots of food with efficient water management systems in place.

In 1820, when Calvinists from New England arrived, they had other ideas about the sharing of food and resources. As the first Christian missionaries in Hawai'i, along with their new god Jehovah, they brought beliefs about private ownership and the making of money through buying and selling of goods, including food.

In 1838, William Richards, one of the leading Calvinists, met all summer with the Mō'ī, Kamehameha III, and the 'Aha Ali'i Nui (Council of High Chiefs) to teach them about the Western world. He taught that the secret paths to Western mana (power) were Christianity, constitutional law, and capitalism. The Mō'ī and the 'Aha Ali'i Nui were amenable to the first two paths. They had already converted to Christianity because Jehovah was deemed to be the foreign god who could save the Hawaiian people from deadly foreign diseases, such as typhoid, smallpox, measles, whooping cough, mumps, tuberculosis, and venereal disease, recently introduced into Hawai'i. And not long after Richards' lectures, the Western-style constitutions were written. But the older Ali'i Nui refused to accept private ownership of land for another ten years, warning that it would lead to the loss of Hawaiian sovereignty. The commoners agreed with them (Kame'eleihiwa 1992).

Their resistance indicates that Native Hawaiians recognized the value of communal use of land to produce food and to prevent hunger. They were worried about how the private ownership of land would affect communal food production. However, the Mō'ī and the Ali'i Nui were forced under foreign military pressure to succumb to the Calvinist advice. In 1843, the French threatened an invasion of Hawai'i, and the American Calvinists urged private ownership of land as the only system that the conquering army would respect. Otherwise, they would conclude that Native Hawaiians owned no land, and they could take all land with impunity. Such was the case in Australia. Furthermore, by 1846, the American Board of Commissioners for Foreign Missions (ABCFM) in New England had notified the American Calvinists that they could no longer financially support their mission in Hawai'i since it had already been accomplished—99 percent of Hawaiians were Christians. Thus the Calvinist missionaries had to choose between leaving Hawai'i for other non-Christian countries in need of missionary activities and finding another nonmissionary occupation. Most chose to become capitalist businessmen, bankers, and sugarcane planters; and in order to succeed in business they needed private ownership of land. Besides American businessmen, other foreigners such as the British and French were also petitioning for private ownership of land as a common practice in "civilized" countries (Kame'eleihiwa 1992).

In addition, the Hawaiian population had dropped precipitously from one million in 1778 to 88,000 in 1846 as a result of foreign diseases. The Calvinist missionaries argued that the only way to save the Hawaiian people from massive depopulation was through private ownership of land. They argued that Hawaiians were dying from venereal diseases because of their "lascivious" behavior. If they were given the land to own, they would be too busy making money to have time to be lascivious (Wyllie in 1846, quoted in Kame'eleihiwa 1992). With the older dissenting Ali'i Nui having passed away, and the Mō'ī and 'Aha Ali'i Nui being desperate to find a solution to the problem of depopulation, this odd Calvinist solution led to King Kamehameha III establishing the Land Commission to Quiet Land Titles in 1846, with William Richards, a Calvinist missionary, in charge of the privatization of land, known as the Māhele (Kame'eleihiwa 1992).

In two years, the process of the 1848 Māhele (division of lands) was well under way. The Māhele, and its subsequent privatization of lands, resulted in a massive disenfranchisement of Native Hawaiians. While the Calvinist missionaries and their sons (their daughters were deemed unfit for land ownership) were allowed 560 acres of land apiece, most commoners who went through the Land Commission received about 3 acres each in land awards called kuleana. By 1893, most of the lands of Hawai'i were owned or controlled by foreigners (Kame'eleihiwa 1992; Trask 1999).

The Māhele also decimated the ancestral Ahupua'a system. On every island, agribusiness, in the form of powerful sugarcane plantations, diverted water away from the communally farmed taro fields, through grand aqueduct schemes taking water from one Ahupua'a to another, and having great adverse impact on Hawaiian kalo farmers (Sproat 2009). With no water, many of the great networks of Lo'i Kalo fields went dry and were abandoned.

On O'ahu the most famous case of water theft was at Wai'āhole in Ko'olaupoko, where water was diverted under the mountain through a series of tunnels in 1912, to the ʻEwa plain for the irrigation of sugarcane plantations by the Waiahole Water Company, a subsidiary of Amfac (Sproat 2009). The taking of Waiʻahole's water caused great misery for Hawaiians, because it took away water from their Loʻi Kalo, crippling the cultivation of kalo. It was not until 1973, with the landmark case of *McBryde Sugar Company v. Robinson*, that water was reaffirmed as a public trust rather than a privately owned commodity (Sproat 2009).

The 1893 overthrow and the 1900 illegal annexation of the Kingdom of Hawaiʻi as a territory of America (UN Treaty Study 1997) exacerbated the demise of the Ahupuaʻa system. Ben Dillingham, an American founder of the Oʻahu Railway and Land Company, built his railroad on stilts through the fishponds along the south shore of Oʻahu, filling them in whenever possible along the way. His dredging of the Ala Wai canal in Waikīkī and the forced filling of Loʻi Kalo and fishponds in that area created an ecological disaster for the Hawaiian people who ate from those lands and ponds. Since Native Hawaiian landowners could not afford to pay for the dredge required by the territorial government to fill in their "mosquito ridden" Loʻi Kalo and fishponds, their lands were confiscated and sold at auction to rich foreigners (Nakamura 1979). Subsequently, after the lands were filled in with Dillingham's dredge, Lorrin Thurston, grandson of one of the first Calvinist missionaries to Hawaiʻi and architect of the 1893 overthrow, became the father of Hawaiʻi tourism, inviting the world to come to Waikīkī to enjoy the beaches and the aloha spirit.

By 1930, most of the fifty-eight fishponds in the south shore districts of Kona and ʻEwa on Oʻahu, comprising 2,280 acres, were filled in and no longer used for the efficient production of fish protein. Still, even with the Loʻi Kalo system in disarray and many of the fishponds destroyed, there was land to grow food, and people ate what they had. The Hawaiians even managed to feed the strangers who walked by their house, as the ancestors had taught them to do with "Hui, e hele mai ʻai!" (Hello, come and eat!).

Furthermore, the 1959 Statehood Act had significant impacts on Native Hawaiian access to land, and ancestral water and land management. When Hawaiʻi became a state, American and foreign investments in Hawaiʻi boomed and the cost of land skyrocketed. Wealthy non-Hawaiians bought up the land and Native Hawaiians had difficulty paying the higher taxes as a result of land speculation. Lack of access to land sharply curtailed the practice of Loʻi Kalo and fishponds.

Personal Reflections

Let me talk about my family's story as an example of how colonialism impacted the native food system and deprived Native Hawaiians of the means to produce their food. In 1961, when I was an eight-year-old child, my family moved to Kahana Valley, in the Moku of Koʻolauloa, on the windward side of Oʻahu. We had been living in Chinatown in urban Honolulu, and my mother, Kathryne Leilani Lee McMahan, wanted to move us out to a healthier environment in the country, where we could escape the car exhaust and eat from the land. My mother worked as a hairdresser, and my stepfather was a sometime carpenter; later I would learn that we were the "working class," but at the time I just thought we were the "working poor." By the time I was eight years old, we had moved nine times, always looking for cheaper rent. We found the cheapest rent possible in Kahana, where we lived for five years.

I remember that even in poverty, food was shared with generosity. When Hawaiians saw you going by, the first thing they would do is to call out, "Hui, e hele mai ʻai!" (Hello, come and eat!), and you would go into their house to eat. It did not matter whether the people of the house did not know you, or had very little food themselves; they would share whatever they had with you. One ate whatever one was offered. Perhaps they only had a little paʻakai (ocean salt) and poi (pounded taro); they would offer all they had to you (Kamakau 1976).

The giving of food was a common practice. For instance, our house was one of the first houses in the valley to have an indoor flush toilet. As people came over to use the toilet, they would usually bring some kind of food as hoʻokupu (gift offering) to the house. In traditional Hawaiian customs, one never went to anyone's house without bringing hoʻokupu as a sign of respect for the house.

Although Kahana once had many acres of Loʻi Kalo and an extensive ʻauwai (water channel) system, by the time my family moved there, the area was overgrown by California grass. It is not clear exactly when those Loʻi Kalo were abandoned, but it may have been after the last Konohiki of Kahana, Sam Pua Haʻaheo, died in the 1950s. Without a Konohiki to manage the flow of water, and to organize monthly volunteer labor to clean the ʻauwai and plant and harvest the loʻi, it is likely that the fields went into disuse. My family loved to eat poi, but we no longer ate it every day. We only bought it when it was on sale, or at Hawaiian parties.

Despite the deterioration of Lo'i Kalo, Native Hawaiians still tried to grow their own food. My family tried to raise as much food as possible by ourselves. My mother never wanted to spend a penny at a grocery store. She felt a sense of security in growing her own food. We had a small yard, where we raised pigs, chickens, turkeys, ducks, and rabbits. We also had a couple of cows staked in an open field down the road. After the fishermen caught akule in the Kahana bay, they would send fish to all of the houses in the valley, so we often had fresh fish to eat. The other source of our protein came from our backyard. We did not grow many vegetables as our backyard was not big enough. However, we did grow string beans and snow peas along the fence, and we had a clump of banana trees next to the pigpen. Most vegetables could be bought cheaply from local farmers, or one could barter duck and chicken eggs instead of using money. We rarely ate lettuces or salads, but we ate a lot of taro leaves, sweet potatoes, tomatoes, and various Chinese cabbages grown locally. Mother regularly sent me and my younger brother to pick breadfruits, mountain apples, mangoes, avocados, and guavas that grew wild in the upper part of the valley. She often made guava jam, especially at Christmas time to give away.

The link that Native Hawaiians tried to keep with land and food was also tenuous. My family left Kahana Valley because the state took it over in 1966, confiscating the land from the Bishop Estate, the Mary Foster Estate, and a few kuleana (native tenant) landowners. Claiming that they intended to make a state park for all the residents of O'ahu to enjoy, they put the residents of Kahana Valley, 99 percent of them Hawaiian, on a month-to-month lease, so that we could be evicted at any time upon short notice.

The story of my family is emblematic of the continuing struggle of Native Hawaiians after the onset of the colonial decimation of the traditional land and water management system. The privatization of land ownership made possible by the 1848 Māhele pushed many Native Hawaiians off their land, and the subsequent American colonization, urbanization, and tourism development further alienated them, depriving them of access to land. The influx of international capital, especially money from Japanese nationals beginning in the 1980s, further increased the cost of land, making it out of reach for most Native Hawaiians. But at the same time, the spirit of communal sharing of food and the joy Native Hawaiians had in being connected with land were not completely lost. Food was critical in keeping Hawaiian traditions and the sense of community.

Thoughts on Food Democracy from Native Hawaiian Perspectives

The centrality of food to the culture of Native Hawaiians has been increasingly recognized among grass-roots activists, and the awareness of the importance of historic food system as described above has only helped to justify their efforts. There are various projects through which people are trying to learn from the ancient food system and to restore some of its components in Hawai'i.

A number of groups are working on the project of Lo'i restoration. For instance, Kako'o 'Ōiwi led by Kanekoa Schultz is a nonprofit organization dedicated to restoring one hundred acres of Lo'i Kalo in Hoi, He'e'ia. Another group, Papahana o Kuaola Hui Ku Maoli Ola founded by Kapalikū Schirman and Rick Barboza, former graduates of Kamakakūokalani Center for Hawaiian Studies (KCHS), is also working to restore twelve acres of the Waipao watershed, an 'ili in He'e'ia, including an extensive Lo'i Kalo.

Fishpond restoration is also under way in many parts of the islands. Kamehameha Schools is restoring the ten-acre Loko Eo fishpond, as well as Kūpopolo Heiau, in partnership with the University of Hawai'i. Paepae o He'eia led by Hi'ilei Kawelo is restoring the eighty-eight-acre He'eia fishpond owned by Kamehameha Schools (Kawelo, this volume). Another KCHS student, Andre Perez, founded a farm in Wai'awa, which is dedicated to making food for people and cleaning up of the land and water contaminated by the military.

Some are examining the potentials of Hawaiian medicinal plants. For instance, Puni Freitas and Doug Hamasaki of Ho'oulu 'Āina and Kokua Kalihi Valley have leased one hundred acres in Kalihi uka to farm forests, Hawaiian medicinal plants, and food in a communal manner. Another example of connecting farming and cultural practices is MA'O Farm that grows seven acres of lettuces and other high-end vegetables, where young Hawaiians work on the farm and also take college-level courses and work towards their degrees.

Because of the limitation of space, I can only mention a few examples of these Native Hawaiian food-related projects. In lieu of a conclusion, I would like to share the recommendations made by the students who participated in the project called Ancestral Visions of 'Āina (land) or AVA Konohiki (see their work at www.AVAkonohiki.com). Many students from the University of Hawai'i's Hawaiian Studies program participated in this research project, which examined the traditional land and water management system and the ways to apply it in today's context. Comparing the food systems before contact and after, the students summarized their recommendations on how to improve the current food system in Hawai'i (Box 1). They are suggestive of the ideas shared by a broader group of Native Hawaiian food activists.

BOX 1. ANCESTRAL VISIONS OF 'ĀINA OR AVA KONOHIKI RECOMMENDATIONS

1. **Have everyone eat from their own Ahupua'a, and grow all of their food needs from their own Ahupua'a.**
   If everyone would look towards the mountain area of each Ahupua'a, they would find ample land for the growing of food, and they would be more careful about using pesticides and other toxins on the land. This Hawaiian custom of eating only from the Ahupua'a in which one lives began around AD 1480 under the direction of the Ali'i Nui Mā'ilikūkahi, as the population was increasing, and they needed to be more prudent about using land for growing food.
2. **Hui, e hele mai 'ai: Be generous in your sharing food with your neighbors, so that no one goes hungry.**
   Plant a surplus to just feed people, not to sell for profit. This ancient Hawaiian custom is based on the idea that plants are happy to be eaten by those who are generous, and that happy plants grow more abundantly to feed people.
3. **'Ai i ka mea loa'a: Eat what there is and grow what we can eat.**
   The traditional Hawaiian ancestral diet was healthy in the extreme. We could not do better. Hawaiian elders believed in eating what was provided, and not to be hard to please. The idea that we should desire food that can only come from 2,500 miles away is nonsensical.
4. **Kanu ka mea'ai Hawai'i: Plant the food that traditionally grows easily in Hawai'i, as well as other healthy vegetables.**
   The food that the ancestors grew traditionally actually represented various Akua (gods or elements) that gave us knowledge

along with health, in a manner similar to the Catholic Communion where people eat the body of Christ to be one with him, and to learn to behave like him with compassion. Remove all genetically modified organism (GMO) food from Hawai'i.

5. **Kanu ka lā'au Lapa'au Hawai'i: Plant Hawaiian medicinal plants that traditionally grew in Hawai'i, so that we are no longer dependent on foreign medicines.**
   Many Hawaiian medicinal plants are not found anywhere else in the world, and we should cherish them. Hawaiians believe that the plants of the earth, tended with aloha, are much better healers of our bodies than synthetic drugs. Moreover, if we do not continue to use Hawaiian medicinal plants, their use will be forgotten, instead of being shared with the world.
6. **Fishpond miracles: Restore remaining fishponds for efficient protein production, and clean the uplands above them of any chemicals.**
   Since traditional Hawaiian fishponds were one of the most efficient means of producing protein, while living in harmony with the land, we should continue those practices forever.
7. **Make the recharge of the aquifer a priority of every Ahupua'a. Remove the concrete culverts from O'ahu streams so that they can recharge the aquifer.**
   Use the uplands of every valley for Lo'i Kalo, which will increase the recharge of the aquifer. The more we can spread the water over the land, the more we will increase water filtration into the aquifer. That is the beauty of the Lo'i Kalo and fishpond networks.

The AVA Konohiki and other projects mentioned above are indicative of the increasingly active role of Native Hawaiians in the local alternative agro-food movement. Or perhaps it is the other way around: For Hawaiians, the growing of food and the feeding of people is another way of walking in the footsteps of the ancestors, and we have always sought some way or another to continue connecting with our land and water, and with our ancestors, through food. Today, the ancestral idea has caught on among non-Hawaiians as well. No doubt there is a synergy of great ideas about food democracy. To recognize the colonial roots in the industrial food system and to learn from the Native Hawaiian ancestral wisdom of sustainable food system management are the first steps in the move towards food democracy.

Notes

1. It is interesting to note that while similar large districts comprised of river valleys were found in other parts of Polynesia, the term *Ahupua'a*, literally "Pig Altar," and the building of such an altar on every boundary of Ahupua'a were only found in Hawai'i (Malo 1959; Henry 1928; Kame'eleihiwa 2009).

2. Again it is significant that Kamapua'a, a god of wetland taro farmers, as well as Lonoikamakahiki, god of yearly inspections of water systems, are first known as in O'ahu traditions, because both are intimately connected with advancements in the increased production of food through the efficient use of water. And, like the Lonoikamakahiki ceremonies that began on O'ahu, Kamapua'a, who was born to a human mother as a divine being in pig form on O'ahu, is only known in the Hawaiian archipelago as a god, and nowhere else in Polynesia (Kame'eleihiwa 1996). This is perhaps because large-scale wetland kalo gardening, and the Lonoikamakahiki ceremonies wherein Kamapua'a was worshipped, are not found elsewhere in Polynesia either.

3. Māui was the only island that had twelve Moku, which is very unusual.

References

Apple, Russell Anderson, and William Kenji Kikuchi. 1975. *Ancient Hawaii Shore Zone Fishponds: An Evaluation of Survivors for Historical Preservation.* Honolulu: National Park Service.

Bishop, S. E. Survey, and Map of Waikiki in 1881. Hawaiian Government Survey. Scale 1:2400. Reg. Map 1398, Hawaii State Survey Office, Honolulu, Hawaii.

Cobb, John Nathan. 1904. "The Commercial Fisheries of the Hawaiian Islands in 1903." In Appendix to the *Report of the Commissioner of Fisheries to the Secretary of Commerce and Labor for the Year Ending June 30, 1904*, 433–512. Washington, DC: Government Press.

Conrow, Joan. 2012. "Military Controls 25 Percent of Oahu Land Mass." *Honolulu Weekly.* May 27.

Fornander, Abraham. 1996. *Fornander's Ancient History of the Hawaiian People* (originally published as Volume II of *An Account of the Polynesian Race*). Honolulu: Mutual Publishing.

Handy, E. S. Craighill, Elizabeth Green Handy, and Mary Kawena Pukui. 1995 (1929). *Native Planters in Old Hawai'i, Their Life, Lore and Environment.* Honolulu: Bishop Museum Press.

Henry, Teuira. 1928. *Ancient Tahiti.* Honolulu: Bishop Museum Press.

HING Natural Disaster Preparedness. 2012. PowerPoint briefing by Lieutenant Governor Brian Schatz on May 18, Slide 6.
'I'i, John Papa. 1959. *Fragments of Hawaiian History.* Translated by Mary Kawena Pukui. Honolulu: Bishop Museum Press.
Kamakau, Samuel Manaiakalani. 1961. *Ruling Chiefs of Hawai'i.* Honolulu: Kamehameha Schools Press.
———. 1964. *Ka Po'e Kahiko.* Translated by Mary Kawena Pukui, edited by Dorothy Barrere. Honolulu: Bishop Museum Press.
———. 1976. *Nā hana a ka po'e kahiko.* Honolulu: Bishop Museum Press.
———. 1991. *Nā mo'olelo akKa po'e kahiko.* Honolulu: Bishop Museum Press.
———. 1996. *Ke Kumu Aupuni.* Honolulu: 'Ahahui 'Olelo Hawai'i.
———. 2001. *Ke Aupuni Mo'i.* Honolulu: Kamehameha Schools Press.
Kamakau, Samuel Manaiakalani, with Kekoa Harmon and Pele Suganuma. 2002. "Ka Mo'olelo o Kahahana, Ka Hopena." *Ka Ho'oilina* [The Legacy], Puke (Volume) 1, Helu (Number) 2:304–305.
Kame'eleihiwa, Lilikalā. 1992. *Native Land and Foreign Desires: Pehea Lā E Pono Ai?* Honolulu: Bishop Museum Press.
———. 1996. *He Mo'olelo Ka'ao o Kamapua'a: A Legendary Tradition of Kamapua'a, the Hawaiian Pig-God.* Honolulu: Bishop Museum Press.
———. 1999. *Nā Wāhine Kapu: Divine Hawaiian Women.* Honolulu: 'Aipohaku Press.
———. 2005. "Kumulipo: A Cosmogonic Guide to Decolonization and Indigenization." *International Indigenous Journal of Entrepreneurship, Advancement, Strategy & Education* (WIPCE special edition, Te Wananga o Aotearoa), Vol. 1, Issue 1:119–130.
———. 2009. "Hawai'i-nui-akea Cousins: Ancestral Gods and Bodies of Knowledge Are Treasures for the Descendants." *Te Kaharoa e-Journal* 2:42–63. http://blogs.libr.canterbury.ac.nz/edu.php?itemid=4955.
Keala, Buddy, with James Graydon and Louisa Castro. 2007. *Loko I'a, A Manual of Hawaiian Fishpond Restoration and Management.* Honolulu: College of Tropical Agriculture and Human Resources, University of Hawai'i at Mānoa.
Kelly, Marion. 1989. "Dynamics of Production Intensification in Pre-contact Hawai'i." In *What's New? A Closer Look at the Process of Innovation,* edited by Sander van der Leeuw and Robin Torrence, 82–106. London: Unwin Hyman.
Kotzebue, Otto Von. 1821. *A Voyage of Discovery into the South Seas and Bering's Straits, for the Purpose of Exploring a North-east Passage . . . Undertaken in the Years 1815, 1818 etc.* 3 vols. London: Longman, Hurst, Rots, Orme and Brown.
Laimana, John Kalei. 2010. "Phenomenal Rise to Literacy in Hawai'i: Hawaiian Society Early 19th Century." MA thesis, Kamakakūokalani Center for Hawaiian Studies, University of Hawai'i at Mānoa.
Malo, Davida. 1827. *He Buke No Ka 'Oihana Kula.* Lāhainā: n.p.
———. 1959. *Hawaiian Antiquities.* Translated by Dr. Nathaniel B. Emerson in 1898. Honolulu: Bishop Museum Press.
McAllister, J. G. 1973. *Archaeology of O'ahu.* Bernice P. Bishop Museum Bulletin 104. Honolulu: Bishop Museum Press.

McKenzie, Edith. 1983. *Hawaiian Genealogies.* Vols. I and II. Lā'ie: Institute for Polynesian Studies.

Menzies, A. 1920. Journal of Archibald Menzies kept during his three visits to the Sandwich, or Hawaiian Islands . . . when acting as surgeon and naturalist on board the HMS *Discovery.* In *Hawaii Nei 128 Years Ago,* edited by F. W. Wilson. Honolulu: F. W. Wilson.

Naehu, Hanohano. 2013. Conversation about fishponds, at Keawanui Fishpond, Moloka'i. August 13.

Nakamura, Barry. 1979. "The Story of Waikiki and the 'Reclamation' Project." Unpublished thesis, Master of Arts in History, University of Hawai'i, May.

Nakuina. M. B. 1894. "Ancient Hawaiian Water Rights and Some of the Customs Pertaining to Them." In *Thrum's Hawaiian Annual,* 79–84. Honolulu: T. G. Thrum.

Office of Hawaiian Affairs. 1988. *Mo'okū'auhau, Genealogies: A Treasure and an Inheritance.* Honolulu: Office of Hawaiian Affairs.

Pratt, High Chiefess Elizabeth Keka'ani'auikalani Kalaninuiohilaukapu. 1920. *History of Keoua Kalanikupu-apā-i-kalani-nui, Father of Hawai'i Kings, and His Descendants, with Notes on Kamehameha I, First King of All Hawai'i.* Honolulu: Honolulu Star-Bulletin.

Ritte, Walter. 2013. Personal communication about fishponds. August 13.

Sproat, D. Kapua'ala. 2009. *Ola I Ka Wai: A Legal Primer for Water Use and Management in Hawai'i.* Honolulu: Ka Huli Ao Center for Excellence in Native Hawaiian Law, University of Hawai'i at Mānoa.

Stannard, David E. 1990. *Before the Horror: The Population of Hawai'i on the Eve of Western Contact.* Honolulu: Social Science Research Institute, University of Hawai'i.

Sterling, Elspeth, and Catherine C. Summers. 1978. *Sites of O'ahu.* Honolulu: Bernice P. Bishop Museum.

Stewart, C. S. 1828. *Journal of a Residence in the Sandwich Islands during the Years 1823, 1824 and 1825.* London: H. Fisher and Son.

Summers, Catherine. 1971. *Moloka'i, a Site Survey.* Honolulu: Bishop Museum PAR Bulletin number 14.

Thompson, Laura. 2013. Personal communication about fishponds on O'ahu. August 5.

Trask, Haunai-Kay. 1999. *From a Native Daughter: Colonialism and Sovereignty in Hawai'i.* Honolulu: University of Hawai'i Press.

Twigg-Smith, Thurston. 1998. *Hawaiian Sovereignty, Do the Facts Really Matter?* Honolulu: Goodale Publishing.

United Nations' Study on Treaties, Agreements and Other Constructive Arrangements between States and Indigenous Populations. 1997. Report to the UN for the "Working Group on Indigenous Peoples." July. Final report by Mr. Miguel Alfonso Martinez Special Rappoteur.

Vancouver, George. 1798. *Voyage of Discovery to the North Pacific Ocean, and Round the World in the Years 1791–95.* Vols. 1, 2, and 3. London: Hakluyt Society.

HI'ILEI KAWELO

Hi'ilei Kawelo is one of the founding members of the nonprofit, Paepae o He'eia, and she still works there as director. Paepae o He'eia is an organization that tries to restore fishponds in He'eia, O'ahu. This narrative is based on an interview conducted by Monique Mironesco in 2012.

Can you talk a little bit about your background? What is your family like?

Growing up in Kahalu'u, I caught, processed, distributed, and cooked fish starting at an early age. Lū'au preparation was important in my family. About 85 percent of the food consumed at family lū'au was locally grown or harvested. We traded fish with cousins or uncles up the road further up in the ahupua'a for kalo, sweet potato, 'ulu or pig. Coming from this background, I was steeped in providing food for my community, especially fish, wana (sea urchin), limu (seaweed), and lobster.

After graduating from Punahou in 1995 and University of Hawai'i at Mānoa (UH Mānoa) in 2001 with a zoology degree, I went to work at the Oceanic Institute in Waimanalo. I learned important skills there, but I wasn't completely satisfied with my job because I wasn't working with the surrounding community. A class in the Hawaiian Studies department at UH Mānoa introduced me to He'eia Fishpond in 2000, and in 2001 I was one of the eight founding members of the nonprofit, Paepae o He'eia. I'm also one of the two remaining original founders still working here.

Can you tell me how Paepae works?

Paepae o He'eia's immediate goal is to provide physical, intellectual, and spiritual sustenance to the nearby community. The long-term goal is to provide food for the community and start reaping benefits from the pond, however long that takes. I'm not sure whether the work I'm doing at He'eia will be able to reverse the ecological problems with the pond spanning decades, but the board members and volunteers have been making progress in that direction. They're focusing on restoring the traditional infrastructure by eradicating invasive mangroves, rebuilding the pond walls and the mākahā (gates). He'eia offers educational programs and engages the community through an outdoor classroom/laboratory site for charter schools. We continue to think of creative ways to sustain the organization through the 'Āina Momona program by charging schools for site visits; harvesting, cleaning, and selling limu; and we grow and sell fish, too. We're also researching whether the pond is a possible site for growing oysters. A healthy pond has a large amount of phytoplankton, and since oysters are filter feeders and they're growing rapidly, that's a good indication that He'eia is a healthy pond. But we cannot meet the Department of Health water quality standards because we have no control over what's going on upstream and in the mountains, or in Kaneohe Bay.

What are some of the challenges you've faced as an organization?

We've been faced with many challenges for Paepae since 2004. Permits to rebuild the wall destroyed by a flood in 1965 have been difficult to obtain from the Army Corps of Engineers. But Paepae o He'eia has learned to use media outlets to make our situation known, and we hope to help other fishpond restoration projects in the process. Whatever we do, we don't want to jeopardize our relationship with the landowner, Kamehameha Schools [KS], since we're lessees to KS. I tell you, "due diligence" are new words in my vocabulary. Once the pond is restored, I'm sure there will continue to be more challenges. Where will we get seed stock for fish to grow in the pond? In order to stock the pond with 'awa and mullet, we need to find a hatchery willing to sell us large amounts of seed stock, but the hatcheries in Hawai'i are struggling. Is there even a market for 'awa and mullet?

People's palates have changed and most of them only want to eat pelagic fish. We continue to try to grow moi, but financially speaking it's a lose money proposition, because moi are carnivorous and fish food is too expensive. People look at us and try to compare Paepae to farms, but it's not a good comparison. There are so many unknowns in a pond—sediment, invasive algae, or other external factors. At a farm, you can see the produce growing from the ground. Here it takes more imagination to "see" what's growing. We want to move to more herbivorous fish. Those fish are more attractive to feeding the kanaka and the elderly people from the Hawaiian and local communities, who like the fishier tasting fish. That's the community we're trying to feed.

How has the project impacted Hawaiʻi's food system?

The community has been eager to buy into the fishpond project because people understand that it's a long-term project. There is at least a three- to five-year timeline in reaching Paepae o Heʻeia's goals. We need to continue working on the mangrove eradication, and restore the wall. [Note: the ceremony celebrating the wall's completion took place in December 2015.] But I'm pretty optimistic about the fishpond's role in Hawaiʻi's food system. The increase in population in ancient Hawaiʻi led the Hawaiians to create 400 fishponds in order to feed more people. This is relevant today because we're in bad shape. The challenge is restoring the traditional agriculture that works *with* the land, *with* the ʻāina, and showing that it can work. To complete the circle, we need more farmers' markets where people can have conversations about what's going on in Hawaiʻi's food system. I talk with Kamehameha Schools [the pond's landowner] about that all the time—the importance of drawing attention to our traditional foods because our people are not healthy. We need to consume food that is made for our people's bodies.

I'm excited to be a producer, to be called a farmer. I'm excited to see how the local food movement is growing. I'm constantly seeking ways to improve the pond and its community outreach by asking how we can improve our volunteer workday lunches, by including more locally grown foods. I reach out to my ʻohana [family] and beyond, and talk to them about eating healthier foods and engaging in sustainable agricultural and consumption practices. If I can convert them, then there is hope for everybody because they are stubborn [laughs].

Paepae o Heʻeia is often compared to MAʻO Farms because people can be fed by their crops. The pond attracts a certain kind of crowd, and doesn't attract another kind of people, but that's OK. At the pond, we are big fans of contributing to one's community by volunteering, and not necessarily donating money, though that's nice, too. We're primarily funded through grants and of course, KS is a large funder. At the moment, we don't generate much revenue, but the pond staff is trying to find innovative ways to engage the community with the local food system. The goal is to model Hawaiian sustainable fishpond practices at Paepae in order to encourage the restoration of other fishponds throughout the island chain. We want to increase the reliance of the population on sustainable, locally grown fish as a viable food source.

4 | Farmers' Markets in Hawai'i

A Local/Global Nexus

Monique Mironesco

From Wal-Mart superstores to farmers' markets, there is a spectrum of food purchasing and distribution options in Hawai'i. This range includes warehouse stores like Costco and Sam's Club; supermarkets like Foodland, Safeway, and Times Supermarket; gourmet stores like Whole Foods; mom and pop stores and health food stores like Down to Earth; food cooperatives like Kokua Market; both large and small "ethnic" markets and convenience stores; community-supported agriculture (CSA) and various types of farmers' markets all over the islands. While national and international supermarket chains dominate the retail markets, the popularity of local food has resulted in the expansion of direct-sale outlets like farmers' markets. The call for localization of food resonates with environmental movements in Hawai'i. Due to Hawai'i's geographic isolation as a chain of islands in the middle of the North Pacific, a large proportion of our food supply comes through shipping and air cargo. While food purchased on the US mainland travels an average of 1,500 miles, Hawai'i's food travels an average of 2,500 miles (Leung and Loke 2008).

This chapter considers farmers' markets as alternative food institutions in Hawai'i. It uses farmers' markets as a lens through which we view the special location (both geographic and metaphorical) of Hawai'i. As a postcolonial state, a generally poor and isolated indigenous population coexists with large numbers of visitors to the islands and divergent settler populations in Hawai'i, each with particular food interests. Finally, it examines the relationship between consumers and producers/vendors within the variety of the Hawai'i farmers' markets in order to provide clarity on how farmers' markets can potentially serve as a political tool to address agricultural issues in Hawai'i.

In examining the variations of farmers' markets and their accomplishments, this chapter critically assesses their role in the pursuit of food democracy. While farmers' markets are often celebrated as a step towards improving the local food system, their contributions to farm security, food security, and consumer awareness and mobilization also vary considerably. The case from Hawai'i highlights the conflicts in farmers' markets such as the tension between improving farmers' income and serving low-income communities, being a space for local community or catering to well-paying tourists. The chapter ends with a discussion of how to address some of these tensions.

Farmers' Markets as Alternative Food Institutions

Farmers' markets have long been an outlet for local food distribution, but their popularity has increased dramatically in recent years. According to the US Department of Agriculture website, there were 8,284 farmers' markets in the United States in August 2014 (http://www.ers.usda.gov). In 1994, when the Agricultural Marketing Service first started to keep track of farmers' markets in the United States, there were 1,755 farmers' markets. The 2014 figure represents a 472 percent increase since tracking began in the United States. Anthropologist Lisa Markowitz argues that farmers' markets "are the 'flagship' of civic agriculture" (Markowitz 2010, 67) given that they are one of the most visible aspects of social change in terms of attitudes towards food, especially considering their meteoric rise in numbers.

Localization of food systems via local retail channels such as farmers' markets has been advocated for various reasons. In her article "From Farm to Table: Making the Most of Your Farmers' Market," registered dietitian Sharon Palmer asserts that "[f]armers' markets deliver nutrient-rich, flavorful food harvested a few short hours before you purchase it . . . [In addition,] dollars spent locally with vendors who grow and operate in a local community, benefit a local community" (Palmer 2010, 2). As in this statement, increased nutrition and taste, as well as investment in the local food community are often mentioned as some of the benefits of farmers' markets.

Another benefit associated with farmers' markets and localized circulation of food is its environmental impact. One study of mainland food miles argues that "of all the energy consumed by the food system, only about 20% goes towards production; the remaining 80% is associated with processing, transport, home refrigeration and preparation" (Hill 2008, 2). An additional perception of the benefits of farmers' markets is that the shortening of the food supply chain enables a closer connection between a consumer and a producer of her/his food. As Connel et al. argue:

> The notion of distance is particularly important to local food systems. . . . Distance refers to "processes that are separating people from the sources of their food and replacing diversified and sustainable food system with a global commodified food system" (Kneen 1995, 24). The material problem is that greater distancing means more resources (e.g. energy) are required to produce a calorie of food. The social problem of distance is that people are disconnected; there is an absence of intimate relations between producers and consumers. (Connel, Smithers, and Joseph 2008, 173)

Farmers' markets help to create such "intimate relations between producers and consumers." Yuna Chiffoleau has found that there are five different types of ties among producers, reinforced through weekly farmers' markets: professional, advisory, cooperative, political, and friendship ties (2009, 222). Farmers' markets are a space for people to see each other at least once a week not only to engage in an economic transaction of food selling and purchasing, but also to reinforce, by their actions of coming to the market, the notion that small-scale diversified agriculture has a valuable place in the food system. Indeed, "farmers' markets, with their vibrancy and visibility, not only engender wider public awareness of food and farming but are central to civic efforts to relocalize agrifood systems" (Gillespie et al. 2007, in Markowitz 2010, 77). Farmers' markets can serve as public spaces for the fostering of civic engagement around community food systems.

In contrast to such positive evaluations of local food and farmers' markets, scholars have also noted contradictions and tensions in farmers' markets. First is the issue of class. Farmers' markets, especially in certain areas, can be seen as elitist, since the food found there may be more expensive than those at the regular retail chains. Farmers' market can be described as a "niche market" disguised by "bourgeois ideology," which "gives consumption the appearance of emancipation" (Goodman and Dupuis 2002, 9). Indeed, Allen and Kovach (2000) argue that cheap, industrialized food has actually served to democratize access to food by attenuating class differences.

Second, farmers' markets and food localization tend to privilege farmers rather than the poor who suffer from food insecurity (Guthman et al. 2006, 682). The tension between food security and farm security observed in other chapters in this volume (Suryanata and Lowry, this volume; Kent, this volume) is also manifested in farmers' markets and calls for a more nuanced analysis of their benefits.

Third, while they have the potential to represent a space for the local rebuttal to the globalized world of agribusiness and large-scale food distribution, Holloway et al. argue that it is important to avoid setting up a duality between "local" food institutions such as farmers' markets and "conventional" agriculture when there are so many additional combinations possible of producer–consumer relations (2007, 2). The term "local" becomes shorthand for "alternative" contrasted with global/conventional, while the reality is that they are not mutually exclusive. Local food does not necessarily exist outside of the globalized capitalist system, neither is it necessarily environmentally sustainable, or socially just.

Nonetheless, many remain optimistic with the opportunities presented by farmers' markets. First, the prices at farmers' markets need to be considered in the context of hidden subsidies to produce foods that are cheaper than those found at farmers' markets. The full environmental cost of imported food might not be reflected in the prices that consumers pay at the cash register of supermarkets or warehouse stores. Furthermore, small, diversified farms are more expensive to operate per acre, but protect overall biodiversity, a benefit that is overlooked by large conventional farms (Kremen, Iles, and Bacon 2012). Since small farmers do not have access to large-scale farm subsidies, they are forced to charge consumers the true price of food, rather than the artificially low price supported by tax dollars.

Additionally, more so than conventional retail outlets such as supermarkets, farmers' markets provide a public and potentially political space. Goodman and DuPuis (2002, 13) point out that while "consumer activism may never overturn the capitalist system . . . as a political action, it does wield power to shape the food system." Farmers' markets have the potential to be a "public arena where a . . . mixture of people can encounter one another as neighbors on common ground" (Kramer 2009, 9). Oftentimes consumer agency is unintentional. Consumers do not necessarily set out to "challenge structures of power in food supply but nevertheless contribute to a practical critique of those structures through their actions and discourse" (Holloway et al. 2007, 15). People shopping at farmers' markets are, at least partially, making conscious political decisions to support a certain kind of local agriculture. I will now turn to the data on farmers' markets around O'ahu and discuss their structures.

Methods

This chapter focuses on the O'ahu farmers' markets that provide a variety of market styles, encompassing different values behind market rules. Participant observation, informal interviews, and conversations with farmers, farmers' market organizers/managers, and market attendees were all methods used for this chapter. I attended three different types of markets on O'ahu, one at Koloa, on Kaua'i, and one in Hilo on Hawai'i Island. In total, I attended various farmers' markets more than fifty times, over a period of almost two years. I attended a total of fourteen markets, though some multiple times: I went to the Kapi'olani Community College farmers' market ten times, the Hale'iwa market sixteen times, the Kailua market four times, the Honolulu (Blaisdell) market twice, the Kaka'ako market three times, and the Sunset Beach and Waialua farmers' markets ten and six times respectively at different times of the day in order to observe potential changes in market attendance and participation. I attended the Mililani, Hawai'i Kai, Wai'anae, Pearlridge, Waipahu, Kalihi, and Wahiawa markets once, typically in the morning. I spoke informally with farmers, vendors (not always one and the same), as well as consumers at the various markets. I spent additional time speaking with three market managers: two of them run the Hawai'i Farm Bureau farmers' markets, and one runs the Hale'iwa, Hawai'i Kai, Pearlridge, and Kaka'ako farmers' markets, along with another Kailua market on Sundays, which is not be confused with the Hawai'i Farm Bureau Thursday night Kailua market. I asked them about the structure of various markets, the patterns of attendance of both vendors/farmers and consumers, as well as the goals for their respective markets. I also spoke with several farmers at the Kapi'olani Community College, Kaka'ako, Kailua, and Hale'iwa farmers' markets. The interpreted narrative analysis was provided to the respondents for their feedback in order to ensure accuracy of their comments' meaning. Their responses are woven through the rest of the chapter, as are my observations of the various markets where they can provide examples for specific areas of interest.

Types of Farmers' Markets in Hawai'i

There are four types of farmers' markets in Hawai'i, each with their respective missions, demographics of producers and consumers, and rules relating to what can be bought and sold at these markets. Hawai'i Farm Bureau Federation (HFBF)–sponsored markets, People's Open Markets (POMs), private markets, and what I call "Anything Goes" markets will all be considered here. Table 4.1 highlights some of the general characteristics of each market type, but does not necessarily constitute an exhaustive list of all farmers' markets on O'ahu since new ones crop up and older ones dissipate on a regular basis.

Hawai'i Farm Bureau Federation–Sponsored Markets

There are no "growers only" farmers' markets in Hawai'i. The HFBF markets come close to approximating that ideal by ensuring that the vendors all sell Hawai'i-grown or Hawai'i-made products, but the latter seems to be under some dispute by various actors. The HFBF operates at least five markets on O'ahu (Kapi'olani Community College on Tuesdays and Saturdays, Kailua on Thursdays, Mililani and Hale'iwa on Sundays, Honolulu/Blaisdell on Wednesdays) and two markets on Hawai'i Island (Keauhou and Kino'ole both on Saturdays). Each market serves a different demographic. According to the market organizers, vendors, and consumers at this particular market, the Kapi'olani Community College farmers' market is very crowded and is frequented by locals, mostly white and Asian, as well as large numbers of tourists (see figure 4.1). An internal HFBF survey found that the Kapi'olani Community College farmers' market was attended by about 50 percent locals and 50 percent tourists (Asagi 2011, personal interview). Anecdotal observation of the market on one day at 9 a.m. puts that figure closer to 70 percent tourists and 30 percent local consumers due to a large number of people streaming out of tour buses on the road adjacent to the market. However, looking at the market over an entire day, from the market starting bell at 7 a.m. all the way until the end of the day, reveals that local consumers tend to come earlier and that the demographics of the market are different depending on the time of the day.

The Kailua market is busy as well, and its demographic includes a mix of white, Asian, and Native Hawaiian customers, but not many visitors.

TABLE 4.1 General characteristics of Oʻahu farmers' markets

Market type	*Hawaiʻi Farm Bureau Federation*	*People's Open Markets (POMs)*	*Private*	*"Anything Goes"*
Locations	Kapiʻolani Community College, Kailua, Mililani, Blaisdell	All over the island. POMs move daily and every two hours.	Haleʻiwa, Hawaiʻi Kai, Ala Moana, Kakaʻako, Pearlridge, Kailua	Sunset Beach Elementary School, Windward Mall, Waiʻanae Coast Comprehensive Center, Waialua Sugar Mill
Demographics	Kapiʻolani Community College and Kailua—a mix of tourists and locals. Mostly white and mixed Asian ethnicities, upper income. Mililani and Blaisdell—mostly locals, mixed Asian ethnicities, middle and lower income.	Mostly locals, immigrants, older, lower income.	Haleʻiwa—mostly white, upper income. Hawaiʻi Kai—mostly Asian, upper income. Kakaʻako (formerly Ala Moana)—mostly Asian, upper income. Kailua—mostly white, upper income. Pearlridge—mostly Asian, middle income.	Sunset Beach Elementary School—mostly tourists, mostly white, upper income. Windward Mall—mostly mixed Asian ethnicities, middle income. Waiʻanae Coast Comprehensive Center—mostly mixed Asian ethnicities and Native Hawaiians, lower middle income. Waialua Sugar Mill—mixed Asian ethnicities and some white, middle and lower income.

TABLE 4.1 *(continued)*

Electronic Benefits Transfer (EBT) / Farmers' Market Nutrition Program (FMNP)	Writing grant to implement EBT system—does not accept FMNP	Accepts EBT and FMNP regularly	Writing grant to implement EBT system—has accepted FMNP in the past	Do not accept any form of EBT or FMNP—no plans to do so, except for Wai'anae Coast Comprehensive Center
Year of inception	2003	1973	2009, 2010, 2011, 2012, 2013	2005, 2006, 2008, 2009
No. of visitors	~4,000, 900, 200, 300	~100–200	~1,000, 700, 1,500, 1,000, 700	~100, 250, 100, 100
Average no. of vendors	20–70 depending on location	10–12 depending on location	55–20 depending on location	15

Source: O'ahu farmers' market managers and author's observations of major farmers' markets on O'ahu.

FIGURE 4.1. Visitors standing in a long line at the Kapiʻolani Community College (KCC) Farmers' Market. Photo by Monique Mironesco.

Both the Kapiʻolani Community College and Kailua markets have a high proportion of prepared foods vendors in relation to farm booths, though the numbers vary weekly and some booths feature *both* whole farm products and prepared foods. The Mililani and Honolulu markets are much smaller and attract a mostly local, mixed Asian and most importantly older demographic. At each of these two markets, shoppers come, buy their products, and leave right away. There is no lingering, no sense of community, even though the regulars clearly know each other and "their" farmers. The Sunday Hawaiʻi Farm Bureau Haleʻiwa market is brand-new and relatively small, with offerings from six farms, and five prepared foods/value-added vendors.

The stated purpose of the HFBF is to unite farming families "for the purpose of analyzing problems and formulating action to ensure the future of agriculture thereby promoting the well-being of farming and the State's economy" (http://hfbf.org/our-purpose/). The HFBF tracks land use, tax and water issues, legislative initiatives, as well as helping members with marketing. Farmers' markets are a direct result of this last purpose. Interestingly, the HFBF also helps its members with "commodity groups," or the aggregation of agricultural-like products, which is likely a holdover from its origins in the era of plantation agriculture. Indeed, much of the criticism leveled at the HFBF relating to its mission is its lack of voice for small family farms. For example, the HFBF has used its farmers' markets to set up educational booths that support a controversial political agenda, most notably a pro-genetically modified organism (GMO) agenda, in addition to food safety certification through a centralized distribution center, and the development of larger farms.

The market managers who run the farmers' markets for the HFBF assert that their purpose is to help small, diversified agriculture farms. To that end, they have set up a two-tiered scale for vendor fees: a lower fee for farm booths, and a higher fee for prepared foods vendors. These markets are among the most well-attended of all of the markets, ranging from five hundred to three thousand people depending on the market and the day, especially on O'ahu, by both producers and consumers. The market managers are social media savvy. They use social networking sites such as Facebook, Twitter, and Instagram to attract customers with enticing pictures of what is available at each market before the markets even begin, along with "tip sheets" on the scheduled vendors for each market on their website.

People's Open Markets

POMs were started on O'ahu in 1973 as a government response to the lack of fresh produce in certain areas and the high cost of living in Hawai'i (see figure 4.2). Their website details their mission statement as follows:

1. Provide the opportunity to purchase fresh agricultural, fishery and other food items at low cost.
2. Support the economic viability of diversified agriculture and fishery in Hawaii by providing market sites for local farmers, fishermen, or their representatives to sell their surplus and off-grade produce.
3. Provide focal point areas for residents to socialize. (http://www1.honolulu.gov/parks/programs/pom/index1.htm, February 2011)

FIGURE 4.2. People's Open Market at 'Ewa Beach. Photo by Monique Mironesco.

These markets are sponsored by the city and county of Honolulu, which provides venues (mostly city parking lots) for vendors to come sell their products for a limited amount of time—about an hour and a half to two hours—and then move on to a new location. There are no requirements that the produce sold be locally grown, though some vendors do sell locally grown produce, especially what the POM website deems "ethnic produce," depending on the particular market location. A supervisor of the POM conducts weekly price surveys at various stores to determine price points for the vendors to follow. Thanks to the original mission of POMs of providing fresh food at low cost, many items can be purchased for lower prices than at regular grocery stores. However, a large number of vendors sell repackaged products bought wholesale or at warehouse stores, breaking them down into smaller amounts or selling them individually. One can find apples, russet potatoes, garlic, and perfect peaches/nectarines at all times of the year. These are all products that cannot be grown in Hawai'i and therefore cannot possibly be local. Many unaware consumers have no idea that these products come from faraway places, and they assume that they are supporting local agriculture by shopping at farmers' markets, not knowing that the product provenance rules for each type of market are different.

The POM vendors are likely to be older immigrants, as are the consumers. Their farm names are displayed on their vans, along with pricing information, a mandatory notice that receipts will be given upon request, and a sign indicating whether that particular farmer accepts Electronic Benefits Transfer cards, commonly referred to as food stamps. On average, at the three POMs I visited in Waipahu, Kalihi, and Wahiawa, of the fourteen vendors, two consistently accepted Electronic Benefits Transfer cards. There were many seniors, both as vendors and as consumers. They seemed to know each other, but business tended to be brisk and to the point. Of all the consumers at the POMs, not a single one I observed brought a reusable shopping bag. One woman even asked for double bags. This indicates that the demographic shopping at this market is completely different than that of the HFBF or other markets on O'ahu. The consumers here are looking for cheap, familiar, and culturally appropriate whole foods. The market is not a new experience for them. They have been attending for many years. These consumers *do* have the time and space discipline to attend and shop at these markets even though they might be held at inconvenient times for working people. Since many of the consumers are seniors, time may not be a particularly onerous problem (moving markets; middle of the day), leading to POMs' continued success as institutions serving certain communities since 1973.

The markets are set up and broken down very fast; there seems to be a well-established routine to each market. The vendors pull up in their vans, display their products on folding tables, put up a price sheet, and wait for customers. There are no tents here, no booths, and no friendly banter. There is no lingering on either the customer or vendor side. The vendors are located relatively far away from each other, so that the entire experience feels a bit disjointed. Because these markets are mostly located in parking lots, the vendors face each other in a long alley-like pattern, which does not encourage a prolonged market experience. People go to their vendor of choice, and then walk back to their cars or a nearby bus stop. When the market has about twenty minutes left, the vendors quietly start packing up in order to move on to the next location in a timely manner.

Private Markets

The economic realities of small and/or diversified family farms—with their lack of government subsidies—have given rise to many small farms growing dependent on grants and nonprofit status to stay afloat. These small and diversified farms have also adopted farmers' markets as a strategy to sell directly to consumers. Private markets, specifically those run by a for-profit organization called FarmLovers Farmers' Markets, exist on Oʻahu in Haleʻiwa, Hawaiʻi Kai, Kakaʻako (formerly at Ala Moana), Pearlridge in Pearl City, and Kailua. The organization is independent from farmers, and is run from fees paid by farmers and/or vendors for advertising and creation of a fair-style atmosphere such as the Cacao Festival, or the Mango Madness recipe contest during their respective seasons. This new type of market provides a framework for higher volume direct farm sales, thereby enabling farmers to avoid selling farm products for wholesale prices to distributors.

The market managers at these markets also use social media to get their message across in order to increase attendance, using Facebook, Twitter, and Instagram to advertise certain vendors or products, or to invite followers to attend certain specialty events at the markets. At these markets, there is an even mix of farmers and prepared foods vendors, but the managers also allow locally made nonfood products to be sold at the markets. Jewelry, handmade soap, artworks, and clothing booths are all present. These markets are "zero-waste," so there is space allocated to recycling programs sponsored by the market, as well as food donation spaces encouraging sustainability and promoting environmental and social consciousness. Finally, the market managers provide booth space for environmental and political nonprofit organizations such as Defend Oʻahu Coalition or North Shore Community Land Trust to disseminate information and raise understanding about their respective causes, thereby encouraging consumers' political engagement.

The market managers' goals, similar to those of the HFBF markets, are to promote local agriculture as much as possible. These market managers put up tents and picnic tables each week in order to promote a public space for community to develop, as people sit down together to share food or just relax in the shade and get to know their neighbors. The demographics of the market in Hale'iwa are about 50 percent people from the surrounding community, 20 percent visitors from outside Hawai'i, and about 30 percent O'ahu residents from other parts of the island (Suitte 2011, personal interview). Each week the market is jam-packed with people. Market attendees can get their shopping done for the week, as well as socialize with friends and neighbors. Lingering is encouraged in the shade of the tents and trees, as well as occasional family movie nights.

Annie Suitte, one of the two market managers, explained that creating a community of like-minded people who support local products was one of their most important goals. However, she also asserted that in terms of precedence, the farmers' economic well-being was a top priority (see figure 4.3). She explained that they had a long waiting list for vendors, but she would rather give a spot to a person with "ten coconuts for sale from their backyard, than another jewelry maker" (Suitte 2011, personal interview). She also explained that the market was a natural space for making connections among farmers. She shared a story about a vendor working for a farmer who was a noni (*Morinda citrifolia,* a type of fruit) expert, but had no place to share his knowledge. The managers knew of another man who

FIGURE 4.3. Farmer's booth at the Hale'iwa Farmers' Market located at Waimea Valley. Photo by Monique Mironesco.

was in need of a farm manager for a small noni plantation in the area and put the two in contact with each other. The noni plantation started to thrive and the man had a place to live and work doing what he loved (Suitte 2011, personal interview). These types of stories are common among market managers, no matter what type of market they run.

The Hawai'i Kai Farmers' Market struggled with attendance due to its location in a bedroom/"soccer" community. The market was on Saturday, when people are taking their children to soccer, and failed to grow as the Hale'iwa market has. After operating for four years, the market closed down in 2014. Suitte attributes this to people's habits in that particular community that do not include shopping in small amounts, or taking the time to get to know their neighbors. Hawai'i Kai is on the upper end of the socioeconomic ladder on O'ahu, so its market attendees are not necessarily in need of Electronic Benefits Transfer card stations. However, Suitte explained that it was an important goal of their roles as managers to figure out a way to accept Electronic Benefits Transfer cards in the near future. No matter what type of system they decide to adopt, she wanted to be sure that it caused the least amount of burden on the farmers. She asserted that it was more important that they spend time talking to and getting to know their customers than fiddling with an Electronic Benefits Transfer machine and the generators and wires required to run them at each booth. They were looking into some kind of centralized system, whereby they would have a central Electronic Benefits Transfer station, and give out farmers' market script in exchange, to be used at any of their markets. This is a similar option to the one mentioned by the HFBF market managers.

The Kaka'ako market is thriving and is proving to be a solid alternative for local people to the Kapi'olani Community College farmers' market in Honolulu on Saturday mornings. This particular market was relocated from the Ala Moana shopping center parking lot due to construction. While there are definitely fewer market attendees at the Kaka'ako market than at the Kapi'olani Community College market, they are mostly local people shopping for their weekly locally grown groceries. The market has a tent in the middle and places to sit in the shade, but its location in the corner of a mall parking lot does not seem to make it appealing for people to linger. That said, the vendors are knowledgeable about the products they are selling and willing to educate consumers about anything from food safety certification rules pending at the legislature to genetically modified organism (GMO) issues in Hawai'i. Along with the Sunday morning Kailua market operated by the organization FarmLovers Farmers' Markets, the Kaka'ako market is one of the newest markets, but it has already garnered a strong core following among consumers and the market managers are looking forward to its continued growth.

"Anything Goes" Markets

The Waialua Farmers' Cooperative Market provides a perfect example of the "Anything Goes" type of market. The vendors are mostly older immigrants, former plantation workers who now lease small plots of land and grow ethnic crops that they are likely to sell in their own community. The consumers indeed generally come from the same demographic as the farmers. However, at this particular market there are also additional products for sale that are clearly not locally grown. Perfectly round and brightly colored oranges and lemons with a Florida sticker and garlic in purple net bags are all displayed without the least bit of irony next to choi sum, baby bok choy (varieties of *Brassica spp.* common in Asian cooking) and kabocha squash, the latter all grown on the North Shore in the Waialua/Hale'iwa area. The market does not usually include vendors of prepared foods, since the original intent of the market was to provide a sales outlet for laid-off plantation workers in the mid- to late 1990s after the closure of the Waialua Sugar Mill. The farmers' cooperative makes the conscious decision to continue this policy at their market at the same time as there is no policy regarding product provenance. There are Fuji apples from Washington State for sale at this market clearly bought in bulk and repackaged into smaller quantities for sale at a relative discount. Brisk sales are done here, but the market layout does not encourage lingering and people tend to go to "their" booth, do their shopping, and then go back to their cars. The immigrant demographic of both the vendors and consumers at this particular market reinforces the postcolonial and globalized nature of the market, both in its attendees and in the products sold. There is less political intent at this particular market to influence consumer decisions to buy local produce than there is a goal to protect former plantation employees and/or immigrant farmers now living in the area by providing them an outlet for direct sales to consumer, therefore bypassing wholesale distributors.

There are a few other "Anything Goes" markets such as the Windward Mall, Fort Street Mall, and Wai'anae Coast Comprehensive Health Center markets, along with another located at the Sunset Beach Elementary School. Originally, this last market was started to provide the community with fresh produce grown on the North Shore of O'ahu. However, in recent times, the market manager has veered away from providing a venue for local farmers and has placed more emphasis on crafts and prepared foods. His argument is that the farmers' market is not only a place to sell food, but also a small business incubator. He hopes to foster the entrepreneurial spirit in new vendors by making the vendor fees extremely low ($10 per vendor per market) in order to encourage vendor attendance. The demographic of the market consumers tends to be very transient. The market manager argues that market attendees are less likely to purchase produce, therefore he encourages prepared food and other types of vendors to attend. There are more jewelry and craft booths at this particular market than there are produce and prepared food booths combined. The local farmers who used to attend this small "farmers'" market now prefer to attend the Hale'iwa Farmers' Market where they are more likely to do brisk business and make lasting connections with loyal consumers.

Challenges: Prepared Foods vs. Whole Foods, Locals vs. Tourists

Among the four types of farmers' markets discussed here, several challenges can be discerned. One challenge is the enforcement of product provenance regulations, if any, in place. Other challenges are keeping a balance between whole foods and prepared foods, and between local customers and tourists.

First, some farmers' markets sell nonlocal vegetables and fruits. The sales of nonlocal food at farmers' markets have been critiqued by many people, as evident in the comments by a local farmer, Gary Maunakea-Forth of MA'O Farms.

> Hawai'i farmers' markets are really at a crossroads, maybe reflective of the entire food-farmer situation in Hawai'i which is also in flux and at a crossroads. A few years ago [there were] a small handful of truly locally-focused markets. As farmers we made really good money, and consistently sold our inventory out. Now there has been a proliferation of markets, some good, some really weak and a complete farce that they should not even be called a

> "farmers' market" because they sell more prepared foods including imported products than local fruit and vegetables.
>
> In this proliferation of markets I hope that local consumers push hard and ask tough questions so that the markets purvey local products and that food vendors are audited to ensure that they are also using local products. The market organizers need to be held accountable; it should be their *kuleana* [responsibility] to make markets a model for zero waste and fairness. . . . Being a small familial community we have the opportunity to provide a much more unique and authentic market product for both locals and tourists. (Maunakea-Forth 2011, e-mail communication)

The problem of authenticity of food sold at farmers' markets is also complicated by the increase in the sales of processed items. A count of vendor booths at the Kapiʻolani Community College, Kailua, and Haleʻiwa farmers' markets reveals that there are many more prepared foods vendors than produce and whole foods vendors, and that the prepared foods are expensive. The product provenance rules are only loosely enforced for prepared foods, thereby creating an unfair system whereby the whole foods vendors are held to a higher standard of locally grown food than the prepared/value-added foods vendors. One example might include some prepared foods vendors at certain farmers' markets using mostly mainland grown items, in conjunction with a few local ingredients, and claiming to consumers to be locally sourced.

The market organizers I interviewed were aware of the controversy over the prevalence of processed foods but added two caveats to this controversy. The first is that the prepared foods booths create an atmosphere that encourages market attendees to linger, to chat, and to create a community interested in a sustainable food system in Hawaiʻi. It is in the interest of the market managers to increase attendance at various markets, and if prepared and value-added foods are big sellers, then they are more likely to be granted booths at the markets since the managers are the ones who decide which vendors get in and where they are allowed to set up. The second caveat is that thus far, farm security is an important mission to the market managers, the consumers, and of course, the farmers themselves. One of the HFBF market managers argued that increased attendance at the market, no matter what type of foods people are buying, is good for farmers because they get exposure to a wider audience (Robello 2011, personal interview). If farmers are able to produce and sell value-added products, which in turn enables them to invest profits into their respective farms, keeping small family farms afloat, then it may be an important benefit to some farmers that cannot be overlooked.

Various farms have been creative in coming up with processed foods utilizing their farm produce. For instance, Otsuji Farms in Hawai'i Kai has devised an ingenious way to sell off-grade kale by dipping it in tempura batter, deep frying it, adding a delicious sauce, and selling it to a long line of consumers at several farmers' markets around O'ahu. Mark Delventhal of Pupukea Greens sells his papaya seed salad dressing alongside his organic lettuce at every farmers' market he attends. North Shore Cattle Company sells more hamburgers each day at the Kapi'olani Community College farmers' market than they sell raw meat. Jeannie Vanna of Big Wave Tomatoes in Waialua sells pizza slices and fried green tomatoes for $6 per slice and $5 per basket, respectively, to long lines of waiting customers each week at the Kapi'olani Community College, Kailua, Kaka'ako, and Hale'iwa farmers' markets. Processed and value-added foods are an important strategy for farmers to increase their revenue. The products tend to have a longer shelf life (dressings or jams, for example). They may cost more for the consumer, but they enable farmers to sell "off-grade" produce that would otherwise be unacceptable to the consumer as a whole food.

The increase in processed foods at farmers' markets has not occurred without controversy, both in Hawai'i and elsewhere. James Kirwan found that value-added foods are deemed acceptable by farmers themselves, but farmers' markets "purists" see them as devaluing the authenticity of the farmers' market experience because they seem to break the connection between the producer and the consumer (Kirwan 2004, 410). Some market goers decry that there are more prepared foods vendors than farmers at various markets in Hawai'i as well.

Processed foods are also related to another salient tension in farmers' markets in Hawai'i where there are a large number of tourists. Some of the market vendors interviewed, along with six regular attendees, two of whom are chefs interested in promoting locally sourced products, grumbled that some farmers' markets have been co-opted by tourism and have morphed into a food institution to serve the tourist market, selling prepared foods, since visitors are unable to store, and unwilling to cook food in their hotels during their Hawai'i vacations. Apparently, this trend is replicated at other large farmers' markets elsewhere as well. One well-known farmer from San Francisco pulled out of the famous Ferry Plaza Farmers' Market in protest because of the pressure to provide value-added products to tourists (Duane

2009, 83). Some farmers and consumers in Hawai'i assert that their farmers' markets are headed in a similar direction.

Tourism's globalizing influence is controversial to dedicated local farmers' market consumers given that it reifies the notion that the markets are becoming tourist attractions rather than alternative food institutions and/or community spaces for local people. One chef told me that the market he used to attend regularly had got so crowded with tourists that he had to get up early and check the tweets from one of the market managers, who took pictures of the products on offer, each Saturday before the market started. He then rushed down to the market, knowing in advance which vendors he would patronize, got his shopping done early, and went home without having had to brave the rush of tourists. This is in great contrast to earlier experiences when he could enjoy the market before the demographics of the attendees had changed so drastically. He used to like coming to the market and spending his Saturday mornings there. Now he wished that all of the produce and meat vendors would be lined up on one aisle of the market so that he could get in and out of there as quickly as possible, without having to dodge the crowds standing in line at the prepared foods vendor booths catering to tourists (Kenney 2011, personal interview). This shift in the attitude of regular/longtime market attendees towards the market is particularly evident at the Kapi'olani Community College farmers' market, where the number of tourists has greatly increased in recent years. The market managers said that there was no concerted effort on their part, the HFBF, or even the Hawai'i Department of Business, Economic Development and Tourism (DBEDT) to advertise the market as a tourist destination, but it had become a "must-see" destination in many tourist publications and was widely advertised in both airline magazines and online travel guides for Hawai'i.

This kind of tension has also been observed elsewhere. In Seattle, for instance, Thomas Tiemann noted that "something has been lost. The market has become a tourist destination and the experience is staged for the tourists. Though there are still tables for farmers . . . the market is less a place for the old, regular Seattle shoppers looking for regional, seasonal food and easy interaction with growers and other shoppers" (Tiemann 2008, 467). Hawai'i's farmers' markets illustrate various trade-offs between processed foods and whole foods, and between local community and tourists. While value-added processed foods attract more tourists because of the ease of transport, certain value-added products also enable farmers to gain additional name recognition by including permanent labels on jars or packaging, which helps farms brand themselves to their customers and encourage repeat business. Yet the higher prices of prepared foods tend to discourage low-income consumers from attending the markets. Moreover, the high price points of prepared foods encourage vendors to develop even more value-added products, perhaps increasing farm security to a certain extent, but decreasing overall food security in Hawai'i by driving up the prices of foods at farmers' markets.

Processed foods and tourists pose critical challenges to the assumptions of farmers' markets as local institutions and a strategy for improving people's access to healthy food. They are often seen as diminishing the value of farmers' markets as a local space for examining assumptions about product provenance and fostering intimate relationships between producers and consumers.

Farm Security vs. Food Security

The observations on farmers' markets discussed above signal a significant trade-off that comes with the increase in tourists and processed foods. The tension between farm security and food security is discernible through market attendees. While the farmers' markets on all islands tend to be well attended, the demographics of the consumers, especially on O'ahu, tell an interesting tale. At the Kapi'olani Community College, Hale'iwa, and Hawai'i Kai markets, there were virtually no Native Hawaiian consumers indicating a disturbing general trend; a large proportion of Native Hawaiians in Hawai'i do not frequent farmers' markets and are unlikely to purchase fresh, locally grown produce. At some other markets, the patrons are relatively wealthy, white or Asian people, and include tourists from the mainland and other countries, especially Japan. There are few Native Hawaiian consumers, though they are minimally represented as producers/vendors.

Many Native Hawaiians are concentrated in the areas of the islands associated with food deserts. Lisa Markowitz defines food deserts as places with "limited access to shops, high prices, poor quality, and narrow variety of food items, especially fresh produce" (Markowitz 2010, 72). In Wai'anae, on the Leeward side of O'ahu, for example, there are only two grocery stores, spanning a sixteen-mile distance. Neither has particularly enticing offerings of produce. There are also a variety of convenience stores and fast food establishments serving the area with unhealthy, cheaply produced food. There is a farmers' market in Wai'anae, no longer operated by the HFBF due to low attendance, but it is the only "Anything Goes" market to accept both Electronic Benefits Transfer cards and Farmers' Market Nutrition Program (FMNP) coupons on a regular basis. If the point of managing a market is to steadily increase attendance, it is no wonder that the lack of crowds at the Wai'anae farmers' market resulted in its being abandoned as a viable market by the HFBF. The Wai'anae Coast Comprehensive Health Center runs the market as a service to the community in order to promote healthy eating and good nutrition in what is considered to be one of O'ahu's primary food deserts, rather than for profit.

The Native Hawaiians' lack of access to the bounty of farmers' markets in Hawai'i is certainly related to the cost of the products found at most farmers' markets, the types of products sold, as well as the location of the most successful markets. Native Hawaiians tend to be poorer and more isolated geographically, especially on O'ahu. Therefore, their lack of access is related both to the cost and to the location of the most abundant markets in relation to where the largest populations of Native Hawaiians live. The lack of access to fresh food for low-income consumers is highly problematic and has serious health implications for Hawai'i due to the high rates of obesity, heart disease, and diabetes, especially in low-income areas.

Food scholars have observed that farm security tends to trump food security concerns (Guthman et al. 2006, 682). Preventing development and rezoning of agricultural lands is a constant battle in Hawai'i (Suryanata and Lowry, this volume), which might explain why farm security is often prioritized over food security. There are concerns that Hawai'i may end up without any viable agricultural lands, should a global event occur to disrupt the current quantity of food imports. One market manager argues that farm security enables agricultural lands to stay in production, so that farmers are able to meet the demand for local food should it continue to increase (Robello 2011, personal interview). Another market manager told me that "cheap food is how we got into this mess in the first place" and that high prices at farmers' markets mean more dollars directly into the pockets of Hawai'i's embattled farmers (Asagi 2011, personal interview). While their concerns with agricultural preservation and farmer welfare

are important, pricing large segments of the community out of the farmers' markets because the fresh food is unaffordable is socially and politically problematic.

Currently, Hawai'i is one of only five states that do not accept FMNP coupons, though some of the markets had accepted them in the past. This program, a part of the Special Supplemental Nutrition Program for Women, Infants, and Children (WIC), allows participants to purchase uncooked, nutritious, locally grown fruits and vegetables. FMNP participants are granted between $10 and $30 per year (not a large amount, to be sure) to promote awareness and use of farmers' markets (http://www.fns.usda.gov/wic/FMNP/FMNPfaqs.htm). Without Electronic Benefits Transfer cards or FMNP coupons accepted at most of Hawai'i's farmers' markets on a regular basis, Hawai'i is lagging behind the rest of the nation in finding innovative ways to provide access to locally grown fresh fruits and vegetables for low-income consumers. This is a concern for some of the market managers associated with the HFBF and private markets. While they do not currently accept Electronic Benefits Transfer cards, they are applying for grants to enable the HFBF markets to accept them in the near future. In addition, the market managers encourage consumers to donate extra produce that they will not eat to the Hawai'i Food Bank. However, to better balance farm security and food security, farmers' market managers need to make greater and more insistent efforts to find ways to incorporate access to healthy, whole foods for low-income consumers, as well as consistently enforce product provenance rules for both whole foods growers and prepared foods vendors.

No farmers' markets in Hawai'i exist unproblematically as there are complicated issues embedded within the farmers' market spaces. There are various demographic groups that could be served in the different types of markets: visitors, farmers, consumers of various ethnic backgrounds and with divergent economic means. These layers of relations complicate the farmers' market space at the same time as the diversity can be the source of its richness. But the class and racial segregation of farmers' markets needs a particular attention. Even though some argue that "food can be a powerful and equalizing social force" (Kramer 2009, 9), the issues of access to healthy food for low-income consumers remain fundamental to the possible contribution of farmers' markets vis-à-vis food democracy in Hawai'i.

Farmers' Market as a Public Space

Supporting small farmers by providing a means for direct farm sales to consumers for farmers who may not have other sales outlets is certainly an important success of the various markets, no matter the type. The creation of a public/political space for the community to address and discuss its concerns regarding various food issues in Hawai'i is another success of the farmers' markets. Are farmers' markets only outlets for direct farm sales, or does their existence encompass political issues as well? Do community members voting with their food dollars gain a voice in the political process regarding food issues in Hawai'i through their attendance at farmers' markets?

The degree to which farmers' markets function as a public space differs significantly between different markets. For instance, market experience for consumers varies from market to market. Rachel Slocum found that in Minneapolis "[f]ew shoppers . . . charge through the Market intent on getting through it in minimum time; the experience tends to be more exploratory" (Slocum 2008, 859). This is a similar observation to the markets in Hale'iwa, Kaka'ako, and Kailua, and for many at the Kapi'olani Community College farmers' market.

In addition, some people attend markets for the opportunity to have conversations with growers. This familiarity is often mentioned by regular farmers' markets consumers as the part of the experience they appreciate the most (Slocum 2008, 864). In Hawai'i as well, some farmers' markets have successfully become a place for a community beyond simply being a place for economic transaction. At the Kapi'olani Community College, Kaka'ako, Kailua, and Hale'iwa farmers' markets, farmers and consumers are often engaged in prolonged conversations. Sometimes, the consumers ask where the produce is grown. Other times, they ask if the farmer has a certain product. If the product is part of the regular offering but is missing from that day's display, the farmer might explain the reason behind its absence, whether it is simply sold out, or its absence is due to inclement weather or some other farm mishap.

Such interaction forms an important facet of consumer education taking place at farmers' markets. Talking to farmers about the reasons behind the crops they have for sale at farmers' markets helps consumers make the connections between larger environmental issues beyond the control of individual farmers such as local farming patterns, the impact of climate change, and agricultural policies. Farmers' markets are a source of good-tasting, fresh, whole foods, as well as a public space for people to understand the sociopolitical implications of purchasing locally grown food. They are not only an outlet for farmers and vendors to sell their produce, but also a place to make lasting relationships and connections with each other and to get involved in community- and market-related issues.

Shopping at farmers' markets takes a special kind of time and space discipline that many people do not have. Farmers' markets only operate on certain days and at certain times. As Sharon Zukin asks, "Why wait for eggs until Wednesday or Saturday when you can buy eggs at Whole Foods any day of the week until 10 p.m.?" (Zukin 2008, 737). While she is discussing the competition between the new Whole Foods market and the long-running Union Square farmers' market in New York City, her point can be applied to any farmers' market in competition with any grocery store. The same type of competition started when Whole Foods opened next door to the Kailua Farmers' Market. When asked about her reaction to this development, one market manager said "bring it on." She asserted that she knew Whole Foods prices could not compete with farmers' market prices, but continues to believe that the loyal following the market has garnered over the years will continue to patronize farmers' booths at the market instead of opting for the convenient Whole Foods experience (Asagi 2011, personal interview). The active role of consumers shopping at farmers' markets necessarily translates into a limited form of political agency because merely making the choice to patronize farmers' markets requires that definite time and place commitment. The motivation is not necessarily only economic, but entails a larger political loyalty to the farmers' market as a site of civic engagement.

The development of farmers' markets as a public space was perhaps most telling from the community mobilization to save the Haleʻiwa Farmers' Market in 2012. In April 2012, the Department of Transportation issued a "cease and desist" order to the market managers for safety concerns due to the Haleʻiwa market's location at the corner of two highways. This event galvanized the community to act to prevent the market's closure. People who consider themselves completely "apolitical" started going to meetings at the legislature, with the governor, with the head of the Department of Transportation, and with other interested parties. They got involved with the political process solely because of their association with the market. While this group included the farmers' market managers, it also included some farmers, as well as consumers who conceptualized the market as an important community resource that they were unwilling to see dismantled by the state. For a time, the market's fate was uncertain, with other locations proposed and then found to be unsuitable by various parties due to size, parking restrictions, or other barriers. One farmer told me that at one meeting, the head of the Department of Transportation admitted that he had never actually been to the market venue and that his decision to close the market was solely based on maps that illustrated the market's location on a highway, even though it was actually adjacent to the highway and protected from passing cars by metal barriers and other landscape features. This unwillingness by state officials to see the market in person, as well as to see the market as an important community resource, illustrates the disconnection between the view of the market from the state's perspective and the grass-roots element of the farmers' markets. The market has since successfully relocated to Waimea Valley on Thursday evenings and while it has fewer numbers of vendors, especially fresh food (rather than prepared foods) vendors, it seems to be thriving.

In many, though certainly not all, cases in Hawai'i, market attendees (including both producers and consumers) are making conscious decisions to support local agriculture in Hawai'i and the markets provide the political and public space to enact those decisions. As the case of the Hale'iwa Farmers' Market location woes indicates, these decisions can take an overtly political turn, enabling people to become politically active in one's community.

While farmers' markets can be described as one of many retail channels where consumers simply come to shop, they seem to be able to foster relationships that are not simply reducible to economic factors. Farmers' markets can be a place for fostering a community, for educating consumers, and for politicizing their consciousness.

Farmers' Markets and Food Democracy

As Goodman and Dupuis suggest, "it is clear that any attempt to integrate how we 'know food' with how we 'grow food' will require rethinking both production and consumption centered notions of politics" (2002, 15). The ability to touch, feel, smell, and take home Hawai'i farm products from farmers' markets tends to embody the struggles of Hawai'i farms and serve as a prospective call to political actions. Is it possible that the large numbers of visitors at certain markets return home and ask questions about their own food systems after attending one of the farmers' markets in Hawai'i? Does smelling a freshly picked vine ripened tomato at a farm booth lead us to question why the tomatoes available at grocery stores look and taste like cardboard? Does the farmers' market experience lead us to question why our broken food system based on imports from thousands of miles away does not support local, sustainable agriculture? Why we should fix it so that it does? Farmers' markets can be an important institutional avenue for food democracy where consumer awareness and connections between food producers and consumers can be fostered.

This chapter has delineated varying challenges of farmers' markets. There are many possibilities ahead for the various farmers' market institutions in Hawai'i, most notably finding the balance between farm security and food security and trying to find a mechanism where farmers' markets can serve the interests of both local farmers and low-income consumers. Another challenge is to quell the concerns about large numbers of prepared foods vendors and tourists at various markets. As discussed above, this issue is more complex than a passing glance would imply. The third challenge is for famers' markets to move beyond a purely economic space and play a political and educational role, serving as the public space with a sense of community and awareness about the problematic status of the food system in Hawai'i. How might farmers' markets serve as a space to start a conversation about these issues? The various farmers' markets in Hawai'i presented in this chapter have divergent track records in meeting these challenges and turning them into opportunities.

Farmers' markets present opportunities for civic engagement. It is necessary not to focus only on producers and consumers, but to discuss more systemic and policy issues in the debate on local food systems. While farmers' markets are often understood as a strategy for localization, the concept of food democracy cautions such advocacy's emphasis on consumer awareness, because it tends to exonerate the government from any political responsibility regarding the politics of food. Individuals "voting with their food dollars" will fall short of engendering a large-scale difference in the food system in Hawai'i. While the change in shopping patterns may shift the structure of local food system to some degree, such emphasis on consumer behavior places the onus on individual consumers and producers to continue the relationship, relieving the government of any active role in supporting small, diversified agriculture.

The state government has made some efforts to support small-scale agriculture in a number of ways, though without necessarily using farmers' markets as a framework for changing the local food system. Farmers' markets are not regulated by the government since the market managers devise and then enforce the various market rules, yet government policies have a direct effect on the economic viability of the small farms that sell at the markets in the face of continually increasing pressures to rezone agricultural areas for development purposes. For example, the state senate commissioned a feasibility study about procurement of local foods for Hawai'i's school lunches (SB 1179–2009) ostensibly to provide a guaranteed market for Hawai'i's farmers. Additionally, the United States Department of Agriculture (USDA) sent a memo to the institutions participating in the National School Lunch Program that they were mandated to apply geographic preferences in procurements for child nutrition programs (USDA Memo Code SP 30–2008). However, neither of these actions has changed anything in the large-scale food system, since the study found that in actuality, asking the Hawai'i Department of Education (HI DOE) to procure locally grown foods for school lunches was *not* currently feasible due to the substantially higher costs of locally grown foods than that of imported foods from the mainland. Additionally, there was no enforcement capacity built into the USDA mandate, so that memo has thus far gone unheeded. Both local and national governments have remained relatively absent from the discussion about local food systems. There is lip service being paid to the support of small-scale farming in Hawai'i, but the politics and policies have not followed suit due to the ever present budgetary constraints faced by the state legislature. It is not enough to raise awareness and educate customers about the importance of buying local food and sustainable agriculture issues. We must continue to reexamine farmers' markets' position in a broader policy and political framework in order to understand whether they are accomplishing the large-scale goal of changing the food system in Hawai'i from one of continued dependence on food imports and the related costs of increased food miles, enabling producer/consumer relationships to develop and thrive, and understanding that not all markets place emphasis on locally grown farm products.

As Guthman et al. have observed, a "pervasive notion in the alternative agricultural movement . . . 'if people only knew' more about food, they would certainly seek organic, healthy, local food" (Guthman et al. 2006,

678). This is certainly a valid critique of the simplistic notion that education and awareness of consumers will change the food system overnight. However, consumers are not passive. They are interrogating their food sources, getting to know farmers, demanding more locally sourced food at their markets. The farmers' markets are weekly reminders that these are important political actions. However, as political agency develops through the farmers' market venue, it also glosses over the fact that the true food costs found at farmers' markets tend to reinforce the reality that healthy, locally grown food products are out of reach for low-income consumers and fundamental policy change, in addition to changes by individual consumers, is necessary. Farmers' markets are a viable *start* to an alternative food purchasing and distribution system towards more democratic community food systems in Hawai'i.

We may not have tipped the balance towards sustainable farming in Hawai'i yet, but with the proliferation of more and larger farmers' markets, there is a potential for greater consumer consciousness and development of political agency to alter the current course of the agro-food system as it is currently manifested in Hawai'i.

References

Allen, Patricia, and M. Kovach. 2000. "The Capitalist Composition of Organic: The Potential of Markets in Fulfilling the Promise of Organic Agriculture." *Agriculture and Human Values* 17:221–232.

Asagi, Lisa. 2011. Hawai'i Farm Bureau Federation farmers' market manager. Personal conversations at Kapi'olani Community College, Blaisdell, and Kailua farmers' markets. February 5 and 24.

Chiffoleau, Yuna. 2009. "From Politics to Co-operation: The Dynamics of Embeddedness in Alternative Food Supply Chains." *Sociologia Ruralis* 49 (3): 218–235.

Connel, David, John Smithers, and Alun Joseph. April 2008. "Farmers' Markets and the 'Good Food' Value Chain: A Preliminary Study." *Local Environment* 13 (3): 169–185.

Duane, Daniel. March/April 2009. "Foodie Beware." *Mother Jones* 34 (2): 83–84.

Erney, Diana. May 2009. "Farm-Fresh Food Comes to Cities." *Organic Gardening* 56 (4): 1.

Goodman, David. 2004. "Rural Europe Redux? Reflections on Alternative Agro-Food Networks and Paradigm Change." *Sociologia Ruralis* 44 (1): 3–16.

Goodman, David, and Melanie Dupuis. 2002. "Knowing Food and Growing Food: Beyond the Production-Consumption Debate in the Sociology of Agriculture." *Sociologia Ruralis* 42 (1): 5–22.

Guthman, Julie, Amy Morris, and Patricia Allen. 2006. "Squaring Farm Security and Food Security in Two Types of Alternative Food Institutions." *Rural Sociology* 71 (4): 662–684.

Hawai'i Farm Bureau Federation. 2016. "Purpose." http://hfbf.org/our-purpose/. Accessed January 20, 2016.

Hill, Holly. 2008. "Food Miles: Background and Marketing." *National Sustainable Agricultural Information Service.* www.attra.ncat.org. Accessed January 6, 2011.

Hinrichs, Claire, and Kathy Kremer. 2002. "Social Inclusion in a Midwest Local Food System Project." *Journal of Poverty* 6 (1): 65–90.

Holloway, Lewis, Moya Kneafsey, Laura Venn, Rosie Cox, Elizabeth Dowler, and Helena Tuomainen. 2007. "Possible Food Economies: A Methodological Framework for Exploring Food Production–Consumption Relationships." *Sociologia Ruralis* 47 (1): 1–19.

Kenney, Ed. 2011 and 2014. Chef/Owner—Town and (formerly) Downtown restaurant. Personal interview with author at Kapi'olani Community College and Kaka'ako farmers' markets. February 5, 2011 and February 8, 2014.

Kirwan, James. October 2004. "Alternative Strategies in the UK Agro-Food System: Interrogating the Alterity of Farmers' Markets." *Sociologia Ruralis* 44 (4): 395–415.

Kramer, Kyle. October 19, 2009. *America* 201 (10): 9.

Kremen, Claire, Alastair Iles, and Christopher Bacon. 2012. "Diversified Farming Systems: An Agroecological, Systems-Based Alternative to Modern Industrial Agriculture." *Ecology & Society* 17 (4): 288–306.

Lamine, Claire. 2005. "Settling Shared Uncertainties: Local Partnerships between Producers and Consumers." *Sociologia Ruralis* 45 (4): 324–345.

Leung, PingSun, and Matthew Loke. 2008. *Economic Impacts of Increasing Hawai'i's Food Self-Sufficiency.* Honolulu, HI: Cooperative Extension Service: College of Tropical Agriculture and Human Resources, University of Hawai'i at Mānoa.

Little, Jo, Brian Ilbery, and David Watts. 2009. "Gender, Consumption and the Relocalisation of Food: A Research Agenda." *Sociologica Ruralis* 49 (3): 201–217.

Mansell, Darrel. 2007. "Market Day in Firenzuola." *Southwest Review* 92 (2): 294–312.

Markowitz, Lisa. 2010. "Expanding Access and Alternatives: Building Farmers' Markets in Low-Income Communities." *Food and Foodways* 18 (1/2): 66–80.

Maunakea-Forth, Gary. 2011. Owner of MA'O Organic Farm. Personal e-mail communication with author. June 9.

Noguchi, Mark. 2014. Chef of Pili Group. Personal conversation with author at Kaka'ako Farmers' Market. February 8.

Palmer, Sharon. 2010. "From Farm to Table: Making the Most of Your Farmers' Market." *Environmental Nutrition* 33, No. 7 (July): 2–3.

"People's Open Market Program." http://www1.honolulu.gov/parks/programs/pom/index1.htm. Accessed February 24, 2011.

Robello, Kacey. 2011. Hawai'i Farm Bureau Federation farmers' market manager. Personal conversations at the Kapi'olani Community College and Mililani farmers' markets. February 5 and 27.

Senate Bill 1179 SD 2. 2009. "Establishes a Farm to School Policy in Chapter 302A, HRS. Establishes a Farm to School Program in the Department of Education." March 18.

Slocum, Rachel. 2008. "Thinking Race through Corporeal Feminist Theory: Divisions and Intimacies at the Minneapolis Farmers' Market." *Social & Cultural Geography* 9 (8): 849–869.

Suitte, Annie. 2011. "FarmLovers Farmers' Markets" manager. Personal interview with author. April 2.

Tiemann, Thomas K. 2008. "Grower-Only Farmers' Markets: Public Spaces and Third Places." *The Journal of Popular Culture* 41 (3): 467–487.

United States Department of Agriculture. 2008. "Applying Geographic Preferences in Procurements for the Child Nutrition Programs." Memo Code SP 30–2008. July 9.

———. 2010. "Hawai'i Fact Sheet." http://www.ers.usda.gov/statefacts/hi.htm. Accessed February 26, 2011.

———. 2011. "Farmers' Market Nutrition Program." http://www.fns.usda.gov/wic/FMNP/FMNPfaqs.htm. Accessed February 26.

———. 2014. "Farmers' Market Growth." http://www.ams.usda.gov. Accessed March 17.

Zukin, Sharon. 2008. "Consuming Authenticity: From Outposts of Difference to Means of Exclusion." *Cultural Studies* 22 (5): 724–748.

5 | Is the Transgene a Grave?

On the Place of Transgenic Papaya in Food Democracy in Hawai'i

NEAL K. ADOLPH AKATSUKA

On May 25, 2010, Jerry Punzal found 397 of the 500 transgenic[1] papaya trees he grew on his farm in Mililani on the island of O'ahu chopped down. He believed the act was "strictly vandalism" (Jinbo 2011). On the night of June 29, 2010, five weeks later, 13,000 papaya trees were decapitated at the papaya farm leased by Laureto Julian in the Puna district on the Big Island[2]—the largest incident of agricultural vandalism recorded in the history of the island (Kakesako 2010b).[3] Julian speculated that the act might have been done by farmers who were jealous of him (Jinbo 2011). However, three days prior to the incident, Julian reported he harvested the first batch of his genetically modified (GM) "Rainbow" and "SunUp" papayas, which led some (such as his brother, William Julian) to speculate a political connection between his growth of GM papaya varieties and the destruction of his trees—the transgenic fruit of which were left to rot on the ground and on damaged trees (Kakesako 2010a). On July 19, 2011, another ten acres of transgenic papaya trees from three separately owned farms in the Puna district were cut down by unknown vandals (Kubota 2011). While the police investigation on all cases remains open, and thus the motivation behind the destruction of the transgenic papaya trees remains unresolved, the speculated connection between agricultural vandalism and anti-GMO (genetically modified organism) politics circulates with ease in local discourse at a time when GM crops are not only criticized in activist literature, but have also long been under public scrutiny in the state legislature.[4,5]

In this chapter, I am concerned with such moments of scrutiny and destruction in which the inseparable relationship between the technoscientific and the political emerges—moments that suggest the (im)possibility of a place for biotechnology in food democracy in Hawaiʻi. In order to analyze these moments, I will first trace the history of the GM papaya. In doing so, however, I recognize that life forms such as the GM papaya do not simply come into existence and now exist in a vacuum, but rather have and will continue to "become with" many beings, human and nonhuman, in a perpetual and historically situated "dance of relating" (Haraway 2008, 4, 25). Tracing the history of the GM papaya and understanding its particular dance then are also to trace the entangled relationships between the multiple life forms that constitute the agricultural way of life of papaya production in Hawaiʻi: humans, the Papaya Ringspot Virus, aphids, and multiple types of papaya.

I argue that the industrial and scientific discourse about the GM papaya tends to install the fruit in the ethical position as the savior of the local papaya industry and bypass the problematic activist criticisms of biotechnology. This position, however, has become destabilized through a range of ecological, economic, and social criticisms leveled by activists. In these criticisms, GM Rainbow papaya is portrayed as the harbinger of contamination and destruction—of organic and conventional papaya agriculture, of the papaya industry in general, and of consumer rights. Furthermore, activist discourse seems to harbor the desire for, and attempt to legitimate, the destruction of transgenic life. These criticisms raise the question: is the transgene a grave? That is, is the transgene (and by extension transgenic papaya) a grave for papaya agriculture and for the papaya industry in general, and thus should it also be sent to a grave (e.g., by no longer being commercially released and thus produced in Hawaiʻi)? I will argue that this discourse, despite the valuable criticisms it offers, should cause hesitation insofar as it shares a similar mechanism, "aversion-displacement" (Bersani 1987), for the legitimated destruction of contaminated and contaminating non-normative beings as the discourse on sexuality and AIDS. If the transgene is a grave, it should be a grave for an ideal agriculture based on purity as well as for the certainty of the place of biotechnology in the local agricultural way of life. Out of this grave may still arise an encounter with an "alien" (Helmreich 2009, 17) that generates situated and accountable responses to, and thus a place for, biotechnology in food democracy in Hawaiʻi and beyond.

The Dance of Transgenic Papaya

Papaya is believed to be native to southern Mexico and Central America. During the Spanish exploration in the sixteenth century, the plant was taken to countries in the Caribbean and Southeast Asia (Yeh et al. 2007). According to the writings of the Dutch traveler Linschoten in 1598, papaya then spread from the Spanish colonies in the Caribbean (i.e., the Spanish Indies) to Malacca and India. Papaya seeds further spread from Malacca or the Philippine Islands throughout the tropical Pacific and became widely distributed on the Pacific islands by 1800 (Pope 1930; Storey 1941). However, accounts differ as to when and how papaya reached the Hawaiian Islands. One account claims that a Spanish horticulturalist, Don Marin, brought papaya between 1800 and 1823 from the Marquesas Islands when he settled in Hawai'i (Storey 1941; Yee et al. 1970). Another account argues that papaya arrived prior to contact with Europeans (in particular with the English in 1778) based on the existence of Hawaiian names for papaya (he-i and milikane) and the claim that Hawaiians did not ordinarily give Hawaiian names to recently introduced fruit but instead adopted the English name or a slightly modified version (Pope 1930).[6]

The difference in whether a European or Polynesians introduced papaya matters in Hawai'i because not all human agencies in the introduction of organisms are treated identically. Many Polynesian introductions are considered native, while European introductions are often identified as alien, classifications that link to Native Hawaiian politics and the history of colonialism in Hawai'i (Helmreich 2009). The ambiguity that surrounds the introduction of papaya in Hawai'i does not lend itself easily to such a classification scheme and as such reminds us not only how native and alien, nature and culture, are constructed categories continually under negotiation (Helmreich 2009), but also how the place of papaya, including transgenic papaya, in Hawai'i from the very beginning cannot be taken for granted nor resolved indefinitely.

Regardless of its origins, the papaya trees originally introduced into Hawai'i are unlike the trees that are presently grown. A botanist with the Bishop Museum, Gerritt P. Wilder, introduced the present type from Barbados, Jamaica, on October 7, 1911 (Yee et al. 1970). This papaya was named the "Solo" (from the Spanish word for "one" or "alone") in 1919, reportedly by Puerto Rican immigrant workers at the Hawai'i Experiment Station

(C. Gonsalves 2001). By 1936, the Solo papaya had replaced the earlier type of papaya trees and became the only variety commercially grown (Yee et al. 1970).[7]

While the majority of the farms that compose the local papaya industry are currently located on the island of Hawai'i and managed by ethnic Filipino farmers, the industry originally began on the island of O'ahu and was primarily managed by ethnic Japanese, ethnic Portuguese, and Native Hawaiian farmers. The shift to the Puna district on the island of Hawai'i began to occur around 1957 when papaya production in the area exceeded the production on O'ahu (7.3 million pounds grown on 220 acres compared to 6.8 million pounds grown on 280 acres on O'ahu) (C. Gonsalves 2001). Several factors prompted this shift: the urbanization and the subsequent high cost of tillable land on O'ahu, the availability of cheaper land with favorable growing conditions in the Puna district with the closing of sugarcane plantations in the area, and particularly the severe outbreak of the Papaya Ringspot Virus that affected the majority of farms on O'ahu in 1959 (C. Gonsalves 2001; Yee et al. 1970). First reported and named by entomologist Dilworth Jensen in 1949, this virus has been characterized as "the most widespread and damaging virus that infects papaya" (D. Gonsalves 1998, 416) and, more bluntly, as "a killer" (C. Gonsalves, Lee, and D. Gonsalves 2004). The virus spreads from papaya tree to tree via the probing touches of aphids on leaves, and the tree gradually exhibits a range of deforming symptoms such as bulging leaf tissue, leaf discoloration, oily dark spots and streaks along the stem, and ring patterns on stunted and misshapen fruits; infection culminates in death of the tree (C. Gonsalves, Lee, and D. Gonsalves 2004; Hine et al. 1965; Kobayashi 1993). With no chemical cure for the infection, farmers sought to avoid the virus altogether and the vast majority of production shifted from the afflicted areas on O'ahu to the unaffected Puna district.

Due in part to the physical isolation of Puna from the rest of the island of Hawai'i by lava rocks as well as the surveying and eradication of infected papaya trees in the adjacent areas of Hilo and Keaau by the Hawai'i Department of Agriculture (HDOA), the district remained relatively free of the virus until 1992. In early May of that year, despite the efforts to move production and avoid the virus, the HDOA identified the virus in a submitted plant sample from the Pahoa area in Puna, which immediately led to the farmer-approved eradication of 600 plants—some identified as infected, and some merely suspected as such. After growers rejected a University of

Hawai'i proposal to completely eradicate all papaya plants in the area,[8] the HDOA enacted a temporary 180-day emergency declaration of the Papaya Ringspot Virus as a pest subject to state statues for eradication in June. This declaration enabled HDOA workers to enter private properties and eradicate infected papaya trees after notification but without the consent of the farmer or landowner. However, instead of enforced total eradication without consent, the HDOA met with representatives from the University of Hawai'i, the papaya industry, and farmers in July, and had begun a more conservative eradication procedure by September. This procedure involved the immediate destruction of infected trees and adjacent trees, and in the case of the persistence of infection in the area after three weeks, the destruction of all trees within a thirty-foot radius of the original infected tree. The practice of thirty-foot radius eradication was discontinued when the emergency declaration expired in December 1992; by the end of April 1993, 5,486 trees had been destroyed in Puna (Kobayashi 1993).

Nonetheless, the virus continued to spread, and aphids were not the only reason. The continued proliferation of the virus was also facilitated by the lack of a strict adherence by farmers to voluntary eradication of infected trees, and reservoirs of the virus were created when farmers abandoned infected fields. By October 1994, the HDOA stopped marking trees for eradication—the virus was declared to be uncontrollable. Infection spread to nearly all of the papaya trees in the Kapoho area of Puna, a third of the productive papaya area in the district, by the end of 1994. The areas of Pohoiki and Kahuawai were completely infected by 1997. Kalapana, the last area in Puna to become completely infected, had discontinued the elimination of infected trees in September 1997; with it, efforts to contain the virus ceased completely. In 1997, only 27.8 million pounds of papaya were produced, a stark contrast to the 53 million pounds produced in 1992 when the virus was first detected (D. Gonsalves and Ferreira 2003).

While state eradication programs were being carried out, several other control measures for the virus were also being developed. For example, collaborative efforts by Cornell and University of Hawai'i scientists Dennis Gonsalves, Mamoru Ishii, Ryoji Namba, Ronald Mau, and Stephen Ferreira presented the products of their research on cross-protection, a project begun in 1978 in anticipation of the eventual spread of the virus to Puna. The University of Hawai'i's Plant Pathology Department distributed cross-protected papaya seedlings to growers on O'ahu in 1993. These seedlings were purposely infected with a mild strain of the virus in order to afford some protection against the effects of more virulent strains, including more consistent (albeit lower) yields and maintenance of plant sugar content (Ferreira et al. 1993). However, this control method was not widely adopted in part because of the negative effects of the deliberate infection on the papaya, the increased level of management required, and the lack of desire among farmers to infect their trees with any strain of the virus (D. Gonsalves 1998).

Another control measure—the focus of this chapter—was transgenic papaya. Funded by a USDA Section 406 grant program, and spurred by the development of "parasite-derived resistance" in transgenic tobacco by Robert Beachy and his colleagues,[9] a collaboration between Dennis Gonsalves, Richard Manshardt (University of Hawai'i), Maureen Fitch (then a graduate student at the University of Hawai'i), and Jerry Slightom (Upjohn Company) began research in 1986 to create and grow a transgenic papaya. Like transgenic tobacco, papaya would be genetically modified to ensure resistance to the Papaya Ringspot Virus through the insertion of the coat protein gene of the virus into the papaya genome. Dennis Gonsalves cloned the coat protein gene of the virus and, with Slightom, sequenced and modified the gene for use in plants. With the use of the "gene gun" (a device that inserts heavy metal particles coated with genetic material into a cell to modify the original genetic material) recently codeveloped by John Sanford at Cornell University, Fitch began work to insert the coat protein gene into papaya embryos in 1988. She developed a line (55–1) of transgenic red-fleshed "Sunset" Solo papayas resistant to exposures to the virus in the greenhouse setting in 1991, and a field trial began on O'ahu in April 1992. Given the grower preference for yellow-fleshed papaya, Manshardt inbred the transgenic "Sunset" Solo papaya (heterozygous for the virus coat protein gene) to obtain the transgenic red-fleshed SunUp/Sunrise Solo papaya (homozygous for the virus coat protein gene), which was then crossed with the yellow-fleshed Kapoho Solo papaya to obtain a yellow-fleshed F_1 hybrid—the Rainbow papaya. A two-and-a-half-year field trial led by Ferreira on an abandoned farm in Puna from 1995 not only confirmed the resistance of the Rainbow and SunUp/Sunrise varieties to the Papaya Ringspot Virus, but also showed that the transgenic papayas performed better than both the cross-protected papayas and the papayas conventionally bred for tolerance to the virus (D. Gonsalves 1998). After the Animal and Plant Health Inspection Service (APHIS), Environmental Protection Agency (EPA), and Food and Drug Administration (FDA) deregulated the transgenic papayas in September 1997, and the Papaya Administrative Committee (PAC) obtained the licenses for the genes and markers used to produce the transgenic papayas from Monsanto, Asgrow Seed, Cambia Biosystems L.L.C., and the Massachusetts Institute of Technology by April 1998, the Rainbow papaya became commercialized and available.

The GM Rainbow papaya finally made its debut on May 1, 1998, when seeds were distributed free to growers.[10,11] As a result, many abandoned fields were reclaimed and new crops planted by late 1998. Papaya production levels steadily increased over the next few years, from 26.7 million pounds in 1998 to 40 million pounds in 2001 (D. Gonsalves and Ferreira 2003). As time went on, the planting of transgenic papaya has also increased. In 1998, all of the papaya grown on the Big Island, the main site of papaya production in the state, was nontransgenic Kapoho Solo papaya (Tripathi, Suzuki, and D. Gonsalves 2006). By 2000, only 37 percent of all papaya planted in the state was Kapoho Solo papaya, while 42 percent was transgenic (Rainbow) papaya (NASS 2009). In 2009, transgenic papaya accounted for 86 percent of all papaya planted (transgenic Rainbow papaya accounted for 77 percent and transgenic SunUp/Sunrise papaya for 9 percent) and nontransgenic Kapoho Solo papaya for a mere 9 percent (NASS 2009). There is no doubt that transgenic papaya has, like the introduction of the Solo papaya from Barbados or the spread of the Papaya Ringspot Virus on Oʻahu, altered the agricultural way of life vis-à-vis papayas in Hawaiʻi.

Based on this historical framing, it is understandable why the Hawaiʻi Papaya Industry Association considers the Rainbow papaya as a savior in their rendition of the "Rainbow Papaya Story," as seen on their website:

> The Rainbow papaya has given Hawaiʻi's farmers a second chance to be able to continue growing papaya. Prior to Rainbow papaya, the papaya ringspot virus disease had become so widespread that the Hawaiʻi papaya industry was on the verge of extinction. Rainbow papaya offers farmers a choice for effectively producing Hawaiʻi premium papaya in areas where the papaya ringspot virus continues to affect papaya plants. (Hawaiʻi Papaya Industry Association 2009)

Supplemented with details on the website of its creation in the public domain outside the direct influence of biotechnology companies, the narrative of Rainbow papaya as a "second chance" for local papaya production in the shadow of death is a particular kind of industry story. Set in the age of "biological control," an age when the existence of inherent biological limits on human manipulation of the biological cannot be taken for granted, this story inscribes ethics into the transgenic papaya by associating its creation with redemption and salvation (Franklin 2003). As such, the story contributes to the creation of an "ethical bypass" around problematic criticisms of biotechnology (Hayden 2003, 201). Similarly, Lisa Weasel argues, "As it turned out, GM papaya was not just resistant to ringspot. It also appeared to be theoretically immune from many of the criticisms that were dogging other GM products on the market" (2009, 148). Who would condemn the papaya that saved the livelihood of Hawai'i farmers from ruin and a local industry from collapsing, particularly at a historical moment marked by the decline of local sugar and pineapple production as well as a low level of food self-sufficiency in Hawai'i?[12]

Is the Transgene a Grave?

On May 25, 2006, ten figures in white hazmat suits were seen removing papaya from an organic farm in Puna. These volunteer farmers and Greenpeace members cordoned off the area with red tape that read "Caution Biohazard." The papaya trees were contaminated. Yet unlike the contaminated papaya removed by farmers and HDOA workers in the previous decade, these papaya trees were not infected with a virus like the Papaya Ringspot Virus. Rather, they were infected by transgenes, the genetic material from an organism whose genome has been modified to contain the genetic material of an organism of a different species—in this case, the genetic material from the SunUp or Rainbow papaya whose genomes contain the coat protein gene of the Papaya Ringspot Virus. Terri Mulroy, the organic farmer who instigated the involvement of Greenpeace, suggested that this contamination might have been caused by the dispersal of seeds from nearby farms growing transgenic SunUp papayas via birds or the wind. These papayas needed to be removed because, as Honaunau organic farmer Melanie Bondera argued, "Organic farmers could lose their certification and their premium-priced crops could be devalued if genetically modified plants take root on their property" (*Honolulu Star Bulletin* 2006).

This incident highlights an activist response to the "ethical bypass" promoters of transgenic papaya attempted to create. Rather than accept the transgenic papaya as a source of salvation for the papaya industry from a plant viral epidemic, activists asserted that the transgenic papaya deserves to be condemned when it too can be a source of contamination and destruction, particularly of nontransgenic plant life and those committed to ensuring its survival (e.g., organic papaya farmers). Rather than a savior, this response situates transgenic papaya as a harbinger of death. Based on a perspective situated in the particular experiences of some organic farmers and anti-GMO activists (Haraway 1991), this figuration echoes through activist reports released in May 2006 by Greenpeace and in June 2006 by Bondera and Mark Query of Hawai'i SEED (formerly GMO Free Hawai'i). In an assessment of the papaya industry after the introduction of the transgenic papaya in Hawai'i, Greenpeace (2006) argues through statistical analyses that the "health" of the industry has generally declined since the introduction of transgenic papaya.[13] The value of papaya prior to the introduction of transgenic papaya was an average of $1.23 per kilogram in 1997, which fell to 89 cents after the release of transgenic papaya in 1998 when Canada and Japan refused to import transgenic papaya, and averaged 80 cents in 2005 (4). The amount of papaya produced after the release of transgenic papaya may have risen to 23.6 million kilograms (52 million pounds) in 2001, but since then has fallen to 13.6 million kilograms (30 million pounds) in 2005, lower than the production levels in the mid-1990s during the viral epidemic (5). The amount of land harvested for papaya was 1,034 hectares at the onset of the viral epidemic in 1993; this dropped to 858 hectares in 1998, and since then has dropped to 587 hectares in 2005 (5). As a result, Greenpeace (2006, 10) concludes, "GE [genetically engineered] papaya is more devastating than the [Papaya Ringspot] virus—the papaya industry was in better shape during the 'virus attack' than it has been ever since the introduction of GE papaya."[14] This is particularly true for organic farmers, portrayed as small scale family farmers that "protect and develop the soil and biodiversity on the islands," who must pay for the sins of transgenic papaya—"the contamination from the GE papaya threatens to kill off this otherwise successful enterprise unless urgent action is taken to protect them from GE contamination" (8).

The plight of such farmers is taken up in publications by the local anti-GMO activist organization, Hawai'i SEED. In a report on the results of a study on the "contamination" of nontransgenic papaya trees by transgenic papaya trees like those on the organic farm in Puna, Bondera and Query (2006) argue that transgenic contamination of nontransgenic papaya crops is inevitable. This conclusion is based on their evidence of two routes of contamination—the pollen flow of transgenic papaya and the unintended mixture of transgenic seeds into the nontransgenic seed supply—supplemented by observations of the practices of farmers and consumers. For example, farmers of transgenic papaya grow trees other than the recommended hermaphrodite trees which primarily self-pollinate, and backyard gardeners may be unintentionally growing transgenic papaya of any gender (male, female, or hermaphrodite), which contributes to transgenic pollen flow. Like the Greenpeace report, the authors emphasize how such contamination entails the destruction of organic and conventional papaya production—originally nontransgenic papaya contaminated by transgenic DNA cannot be sold to certain lucrative markets that do not accept transgenic papaya (particularly, Japan) (15),[15] reduces the price of the papaya (Puna farmers interviewed in 2003 reported transgenic papaya garnered a price of 13 to 17 cents per pound while nontransgenic [though not necessarily organic] papaya garnered 45 to 75 cents) (7), and forces organic farmers to bear the financial responsibility of ensuring their crops are nontransgenic in accordance with state certification guidelines (e.g., testing papaya trees for transgenic contamination and the destruction of trees that test positive) (3). In a separate Hawai'i SEED publication, Bondera (2006) also notes transgenic papaya (specifically, Rainbow and SunUp) are susceptible to blackspot fungus, which can create additional financial burdens on papaya growers and point to the existence of other agricultural problems beyond just the Papaya Ringspot Virus. A source of contamination and destruction to many, a savior to a select few (if any): "Many farmers are losing money growing GMO Papaya. The GMO Papayas are being dumped or fed to pigs. Some farmers are giving up growing papayas altogether" (Bondera and Query 2006, 7).

Such criticisms of contamination and destruction caused by transgenic papaya not only destabilize the ethical bypass around the concerns about the transgenic nature of Rainbow papaya,[16] but also raise the counter question, is the transgene a grave? The affirmative response of Bondera and Query (2006) seems clear in their recommendation that "considering the adverse consequences of the GMO Papaya, the Hawai'i Department of Agriculture should not commercially release any more GMO crops in Hawai'i." I would argue that this affirmation harbors a prevalent desire in anti-GMO politics for a new relation between society (and by extension, food democracy) and biotechnology based on the destruction of transgenic life. This desire does not necessarily become embodied in the direct acts of crop destruction (indeed, I doubt much of it does). Local activists are heterogeneous and social activism can take multiple forms that do not rely upon direct action. I want to draw attention not to literal destruction, but rather to a desire for the absence of transgenic crops in Hawai'i, an absence that can be achieved in multiple (legal) ways. That is, in this formulation, transgenes are a grave (for organic papaya agriculture and for the papaya industry in general) and thus should also be sent to a grave (e.g., by no longer being commercially released and thus produced in Hawai'i).

On one hand, I readily acknowledge the value of the critical and situated knowledge produced by anti-GMO activists. It is not only unhelpful, but also undemocratic to discount or refuse to take their criticisms seriously, particularly in regard to the limitations of the transgenic papaya as an agricultural solution and the oftentimes unequally distributed consequences of the use of transgenic papaya. The recent allegations of activist involvement in the aforementioned acts of vandalism on papaya farms in 2010 and 2011 that attempt to reduce activist politics to senseless and bigoted destruction are also damaging. Recall that these incidents are still under investigation, and no anti-GMO activist involvement has been proven. However, despite a lack of evidence, several people have made this link. For example, in a *Honolulu Star-Advertiser* (2011a) editorial, the police are called upon to "step up their investigation of this criminality and, along with the public, recognize that this goes beyond mere property damage and is becoming a form of agricultural terrorism" likely connected to the controversial use of transgenic papaya that has resulted in "friction between organic and GE [genetically engineered] papaya farmers." In an article by Alan Gottlieb (former president of the Hawai'i Cattlemen's Council) and Lorie Farrell (executive director of the Big Island Farm Bureau) (2011), cosigned by Myrone Murakami (president of the Hawai'i Farm Bureau Federation) and Rusty Perry (president of the Hawai'i Papaya Growers Association), the authors go further and speculate that the acts of "agricultural terrorism" were perpetrated by "eco-terrorists," "extreme forces" that have unfortunately "made their way to our peaceful shores." While the authors claim they cannot assume the recent destructions of papaya crops were performed by anti-GMO activists, they nonetheless assert the perpetrators were "fueled by the same extreme desire: to advance their narrow beliefs and encroach on the free will of those with whom they disagree [i.e., transgenic papaya growers]" (Gottlieb and Farrell 2011). Such representations result in an ironic refusal to enter into a "dance of relating," a becoming with those who have differentially situated and contrary positions, and thus delimit democratic exchange.

At the same time, I cannot help but hesitate to accede to the particular desire for the destruction of transgenic organisms. Writing about transgenic organisms like the tumor-susceptible laboratory mouse, OncoMouse, Donna J. Haraway (1997) argues that the concerns of activists about contamination, the sanctity of life, and purity of type, should induce anxiety in light of the history and politics of race and immigration, the mixed and the alien, in Europe and the United States. However, while Haraway "cannot hear discussion of disharmonious crosses among organic beings and of implanted alien genes without hearing a racially inflected and xenophobic symphony" (62), I cannot hear yearnings for the destruction of contaminating and contaminated non-normative beings whose emergence has declined the "health" of the papaya industry without hearing echoes from the history and politics of AIDS in the United States about transgressive sexuality, fears of infection from polluted blood supplies and bodies, and a desire to exterminate homosexuals.

In "Is the Rectum a Grave?" Leo Bersani (1987) analyzes such echoes in the discourse of sexuality and AIDS that portray homosexuals as killers (i.e., their promiscuous performance of anal sex makes them the contaminated purveyor of a deadly virus among the public) and legitimize a desire to kill homosexuals. This discourse and the gay responses to it are saturated by what Bersani calls "aversion-displacements" (220)—that is, aversions towards and displacements of a frank discussion of sexual relations in all their corporeal messiness, complexity, and particularity, guided by a "pastoralizing," idealizing, or redemptive project about purity and what sex should be (221). As a result, "the brutality is identical to the idealization" (221).

Like the discourse of sexuality and AIDS, a series of aversion-displacements occur in activist representations of transgenes qua grave that turn away from the messy particularities of the history of and multispecies relationships with the transgenic papaya towards a moralistic discourse about genetic modification and pure forms of agriculture, epitomized in such figures as the organic farmer. For example, activist representations that focus on the transgenic nature of the GM papaya displace the role of other beings, human and nonhuman, in the history of the transgenic papaya and thus tend to obscure the degree of urgency and necessity behind its birth. The Papaya Ringspot Virus ravaged the fields of papaya and became widespread not only because of aphids, but also because of the farmers who did not eradicate infected trees and abandoned infected fields that could become reservoirs of the virus. The viral epidemic and the agricultural way of life which enabled it then spurred on university scientists to develop the transgenic papaya, one of several countermeasures which happened to be the most effective. The virus continues to affect the production of nontransgenic papaya because producers need to cut down infected trees and comply with the Plant Variety Protection (PVP) protocol (C. Gonsalves, Lee, and D. Gonsalves 2004), and there are constraints placed upon where nontransgenic papaya can be grown (i.e., there are limited sites where nontransgenic papayas can flourish and are also isolated from the virus) (D. Gonsalves and Ferreira 2003).[17] According to O'ahu papaya farmer Ken Kamiya, "[f]ew local nongenetically modified papayas reach maturity without developing the ring spot virus" and thus growers "have been reluctant to plant a non-GMO" (Hao 2010, A26). Criticisms of the transgenic papaya need to inherit this history.

Another example is the decline in papaya production and value solely attributed to the introduction of transgenic papaya that obscures other historical contingencies. In 2002, papaya production declined at least in part because of

> fruit scarring, fungal disease, and slower fruit maturation as a result of adverse weather. In addition, the Papaya Administrative Committee (PAC) disbanded in October 2002 . . . this entity was responsible for developing the papaya market, conducting product and market research, and advertising.[18] (Pesante 2003, 1)

Furthermore, the prices for exported papaya to Japan have been falling not only because of the restrictions placed on transgenic papaya, but also because of the general decline in the world price of papaya between 1995 and 2008 (Parcon et al. 2010). There is also increased market competition from papaya production in the Philippines, whose papaya prices are lower (about half the price of Hawaiian papayas), distance from Japan is shorter (which means a decreased chance of damage to exported papaya), and supply of nontransgenic papaya is larger (Pesante 2003). In 2008, the Philippines eclipsed Hawai'i as the largest supplier of papayas to Japan with a market share of 72 percent (Parcon et al. 2010).

Homosexuals and transgenic organisms are thus queer kin in the way they are figured as graves. When Bersani (1987, 204) references Simon Watney to emphasize how gay men in this figuration "are officially regarded, in our entirety, as a disposable constituency" then, I would argue there is also a related significance for the desired destruction of transgenic organisms. Bersani's insight resonates with Giorgio Agamben (1998) who, extending Michel Foucault's (1990) idea of "biopower," argues that life itself, or "bare life" (*zoe*), has come to "occupy the very center of the political scene of modernity" and thus calculations about value and nonvalue. Beyond a certain threshold, which every society decides upon, life ceases to be politically relevant (moves from the realm of *bios* [political and biographical life] to *zoe*) and those identified with such "bare life" can be killed without punishment (Agamben 1998, 139). When Bondera and Query (2006, 17) recommend "considering the adverse consequences of the GMO Papaya, the Hawai'i Department of Agriculture should not commercially release any more GMO crops in Hawai'i," they also place transgenes (and by extension transgenic papaya and other transgenic organisms) in the realm of "bare life," a movement which threatens not only the survival of transgenic papaya, but also another group displaced with the focus on transgenes—farmers who depend on transgenic papaya for their own livelihood. If only because of a commitment to the livelihoods of such farmers, transgenic papayas cannot be treated as solely within the realm of "bare life" and thus killable without consequences. Yet it also cannot be treated as a life form whose survival does not come without consequences for others (such as organic farmers). Aversions towards and displacements of the messiness, complexity, and particularity such engagement with transgenic papaya entails will only lead food democracy itself to a grave.

Planting Seeds for a Food Democracy

The only grave the transgene should be then is a grave for an ideal agriculture based on purity and for certainty of the place of biotechnology in the local agricultural way of life. With this in mind, on December 29, 2010, I planted a few Rainbow papaya seeds purchased from the University of Hawai'i at Mānoa Seed Program in my backyard on O'ahu. Not only did this enter me into a situated relationship with the papaya industry because

I had to sign and agree to the terms of a contract (more specifically, a material transfer and proprietary rights agreement) with the PAC,[19] but also because by planting the seeds I contributed to the saturation of the Hawaiian seedbed with transgenic organisms and thus possible transgenic contamination of nontransgenic plants in the future. However, I do not want to treat this moment as the materialization of the transgene as a grave, a harbinger of contamination and destruction, or as a savior of the papaya industry, and thus contribute to the aversion displacements that already configure the discourse of GM crops. Rather, I argue this moment could be a close encounter of the transgenic kind with an alien—"a life form whose place in our forms of life is yet to be determined" (Helmreich 2009, 17). The planting of a transgenic papaya seed may be an opportunity to touch, become with, and thus form connections of care and responsibility towards other growers, activists, scientists, aphids, papayas, and so on around me that inherit the uneven human and nonhuman relations that have become entangled in the history of the local papaya production (Haraway 2008). This would mean actors (including myself) must move beyond aversion displacements towards frank discussions of GM papaya and how it contributes to and hinders the flourishing of others in the pursuit to grow and eat the kind of foods "we" may desire or need—food democracy.

Such food democracy would require, for example, transgenic papaya farmers to make sure to cut down male transgenic papaya trees that increase transgenic pollen flow and only grow the recommended hermaphrodite trees that self-pollinate (Bondera and Query 2006). It would also require policymakers to support the measures that enable consumers to be more aware of transgenic papaya and thus not plant transgenic papaya seeds if they are not prepared to deal with the responsibility of growing such plants (Bondera 2006). Not everyone will be satisfied with the results of such reconfigured practices, nor will the danger of a reification of existing power relations be avoided, but this process seems much more democratic to me than the strict acceptance or rejection of transgenic life and the actors involved in the agricultural forms of life built around it.

Such a reconfigured relationship towards biotechnology in society between humans and nonhumans (transgenic or otherwise) seems particularly possible and necessary given the decision in April 2010 by Japan, the former primary market for Hawaiian papaya prior to the cultivation of transgenic papaya, to finally accept transgenic papaya for import (Cline

2010). I find the possible beginnings of such a reconfiguration in a conversation with John Takada (a pseudonym), who is involved in the international discussions to open the Japanese market for locally grown transgenic papaya on May 12, 2009.

> I am doing papaya because it is the right thing to do, yeah? And I don't want to circumvent the letter of the law and do something that is not right to the consumer [regarding how to market transgenic Rainbow papaya in Japan in light of the Japanese labeling regulations for transgenic food]. I basically believe the consumer has a right to know . . . I am not doing this because I want to push GMO . . . I'm just doing this because to help the papaya farmer, you know? But in doing so, I become, I walk into this new area. I understand my responsibilities and I don't want to like trod on the rights of consumers.

While not everyone would agree with the work of Takada in Japan, in his entanglement with and feelings of responsibility towards both transgenic papaya farmers in Hawai'i and Japanese consumers, most of whom do not desire to eat transgenic food (Akatsuka 2010), Takada engages in the kind of accountability necessary for biotechnology to have a place, contested though it may be, in food democracy.

Notes

1. In this chapter, I use the terms "genetically modified," "genetically engineered," and "transgenic" interchangeably to denote an organism whose genome has been altered through the insertion of genetic material from an organism of another species.
2. The "Big Island" locally refers to the island of Hawai'i and is used throughout this chapter to disambiguate the island of Hawai'i from the use of "Hawai'i" to refer to all the islands that compose the State of Hawai'i.
3. While Kakesako (2010b) cites the figure of 13,000, other articles cite the total number of destroyed papaya trees at around 8,500 (e.g., *Honolulu Star-Advertiser* 2011b; Jinbo 2011; Kubota 2011).
4. A connection with anti-GMO politics has not been the only suggested reason behind the acts of agricultural vandalism (as indicated by the conjectures of Punzal and Julian). Another suggested cause is the retaliation of game hunters accused of trespassing on farm property (Salazar and Sekiya 2011). My focus on the connection with anti-GMO politics is not to suggest the acts of vandalism are a clear example of such a connection, but rather to demonstrate the salience of such a connection in local discourse and to analyze the consequences of the rhetoric used for food democracy.
5. Hawai'i has been one of the most (if not the most) active states in legislating biotechnology in recent years. Twenty-three bills and resolutions were introduced between 2001 and 2002, twenty-four in 2003, fourteen in 2004, and forty-four between

2005 and 2006 (Pew Initiative on Food and Biotechnology 2005; Pew Initiative on Food and Biotechnology 2007). In 2008, Hawai'i was the second most active state in legislating biotechnology with twenty-two bills (Pechar and Tatum 2008). In the 2011 regular session, nine bills have been introduced so far into the Hawai'i State Legislature at the time of this writing.

6. The history outlined is heavily based on the record of events in Ferreira et al. (1993), D. Gonsalves (1998), D. Gonsalves and Ferreira (2003), C. Gonsalves, Lee, and D. Gonsalves (2004), Isherwood (1994), Kobayashi (1993), Manshardt (1993), Nishina et al. (1998), and Yoon (1999).

7. The various papaya types presently sold in Hawai'i (such as the widely grown "Kapoho" papaya and "Waimanalo" papaya) are all strains bred from the Solo papaya variety (Yee et al. 1970, 6).

8. According to Kobayashi (1993, 3), the university's plan was based on the inability to predict a pattern of how the virus would spread and the unavailability of a method to determine latent infections in plants.

9. In the parasite-derived resistance model, genes from a parasite are genetically engineered into a plant to confer resistance to the detrimental effects of the same or related parasite. In the case of the transgenic tobacco developed by the Beachy group, the coat protein gene of the Tobacco Mosaic Virus was inserted into tobacco and found to confer resistance to the virus (D. Gonsalves and Ferreira 2003).

10. While both Rainbow and SunUp/Sunrise papaya seeds were distributed, the vast majority was of the Rainbow variety (D. Gonsalves and Ferreira 2003). The seeds are no longer distributed for free though—the PAC, composed of growers, decided to charge a fee for seeds in order to raise more seeds for distribution (D. Gonsalves 2003).

11. Based on demographic data gathered in 1999 of active papaya farmers identified in Puna on the Big Island (C. Gonsalves 2001; C. Gonsalves, Lee, and D. Gonsalves 2004), most farmers were male (90 percent); of Filipino ethnicity (91 percent, followed by Japanese ethnicity at 4 percent and Caucasian/other ethnicity at 4 percent); ranged in age from twenty-two to seventy-three years (with an average age of forty-seven years); worked an off-farm job for between 15 and 70 hours per week (46 percent); married and with a spouse who also worked on the farm for 2.5 to 70 hours per week as well as an off-farm job (47 percent); and with an education in either the United States or the Philippines (29 percent with an education up to elementary or high school, 43 percent up to high school, and 28 percent up to college or postgraduate school).

12. According to the Hawai'i Department of Agriculture (2008), 85 to 90 percent of food in Hawai'i is imported.

13. The health of the industry since 2005 is somewhat complicated. According to the National Agricultural Statistics Service (NASS) (2010, 1), the value of papaya has since risen to an average of 99 cents per kilogram (45 cents per pound) in 2009. The amount of papaya produced also rose slightly to 14.3 million kilograms (31.5 million pounds) in 2009. The amount of land harvested for papaya has dropped further to 536 hectares in 2009. While these numbers suggest a rebound in the health of the industry to a certain extent though, the argument of a general decline since the introduction of transgenic papaya still holds.

14. Like any statistical analysis, the trend of decline and hardship noted by the report does not necessarily hold true in all cases. "Some growers commented on the economic difficulties of growing papayas with the high cost of inputs, labor, and low

returns for their crop. Additional concerns were weather conditions and controlling disease and pests in their orchards. Other growers mentioned their orchards were doing well with no major incidences" (NASS 2009, 3).

15. The truth of the claim of the inability to sell nontransgenic papaya contaminated by transgenic pollen as organic is not completely clear. On the one hand, USDA regulations indicate that testing can be required if an organic certifier believes an organic farm has been cross-pollinated by transgenic pollen. According to Luke Anderson (2006, 41–42), if testing confirms cross-pollination, the farmer would lose organic certification. On the other hand, according to Manshardt (2002, 1), papaya farmers will not lose their organic certification due to cross-pollination of their crops: Although the produced papaya may have some transgenic seeds, it "has the same genetic constitution as the tree that produces it. Said another way, if you plant trees that you know are *not* genetically engineered, the edible fruit they will produce will *not* be genetically altered by cross-pollination, no matter what the source of the pollen." Furthermore, he notes that according to USDA regulations, fruit from nontransgenic papaya that has been cross-pollinated can be sold as organic if the grower can document that, except for the cross-pollination, the crop was grown according to organic standards and that efforts were made to avoid cross-pollination (2). To complicate the matters further, the constraints on the ability to sell cross-pollinated nontransgenic papaya may also change in the near future (according to one official in the Hawai'i Department of Agriculture I consulted, by the end of 2011) when transgenic papayas are accepted by Japan for importation. For example, with this acceptance, will papayas with nontransgenic flesh, grown according to organic standards, but cross-pollinated and therefore containing at least some transgenic seeds, count as nontransgenic and organic, nontransgenic but not organic, or transgenic? Will the amount of transgenic seeds matter in how these papayas are classified, labeled, and sold in Japan? How such questions are answered will reconfigure the significance of transgenic cross-pollination and its effect on local papaya farmers.

16. There are other criticisms of biotechnology applicable to the case of the Rainbow papaya. For example, Greenpeace (2003) also raises concerns about the patenting of papaya seeds. While such criticisms are important to consider, they are outside the scope of my paper and left to others to think through. In this paper, I decided to focus on the issues of contamination and destruction—particularly related to the impact of gene flow on organic and conventional papaya agriculture as well as the economic survival of the papaya industry in general—because they seem to be one of the most frequently raised in activist literature, the local media, and recent testimony given at legislative hearings on bills related to GM food and crops. While such a focus comes at the cost of considering a wider scope of criticisms, such criticisms deserve their own substantial discussion in order to do justice to the complexity of the issues they raise.

17. According to D. Gonsalves and Ferreira (2003), "Soon after the discovery of PRSV in Puna in 1992, new plantations were started on different areas of Hawai'i Island where PRSV had not been identified. Although these areas did not have the virus, [the dominant commercially grown and nontransgenic] Kapoho variety did not adapt well to these regions in that the fruit were generally smaller than those grown in Puna. The result was papaya production continued to drop and Hawai'i began to lose market share in mainland USA and it became more difficult to maintain the shipment of quality nontransgenic papaya to Japan."

18. The PAC was formed in 1971 by the Hawai'i papaya industry with the voluntary creation of a Marketing Order (Federal Marketing Order #928) authorized by the United States Congress. The Marketing Order placed the industry under the USDA and made the industry eligible to receive assistance with marketing problems (e.g., advertising, quality and quantity regulations, and standardization of containers) as well as research and development (C. Gonsalves 2001, 26). The PAC was disbanded in 2002 based on the results of a grower referendum and subsequent termination of the Marketing Order. The Hawai'i Papaya Industry Association (HPIA) succeeded the PAC, but there are significant differences between the two organizations. For example, the HPIA, unlike the PAC, does not require membership, submission of production records, or contributions towards research and marketing. Furthermore, unlike the PAC, the HPIA is not part of the USDA and cannot distribute transgenic papaya seeds for free (C. Gonsalves, Lee, and D. Gonsalves 2004).

19. While the PAC was disbanded in 2002, the contract negotiated by the PAC for the licenses of genes and markers used to produce transgenic papayas still holds. Under this contract, the Hawai'i Agricultural Research Center (HARC) continues to produce transgenic papaya seeds and the HPIA distributes the seeds for a fee (C. Gonsalves, Lee, and D. Gonsalves 2004).

References

Agamben, Giorgio. 1998 [1995]. *Homo Sacer: Sovereign Power and Bare Life.* Translated by Daniel Heller-Roazen. Stanford, CA: Stanford University Press.

Akatsuka, Neal. 2010. "The Haunting Erotics of Gastronomic Desire as Bodily Penetration." *Lambda Alpha Journal* 40:3–20.

"Altered Papaya Trees Removed." 2006. *Honolulu Star Bulletin.* May 26. http://archives.starbulletin.com/2006/05/26/news/story10.html.

Anderson, Luke. 2006. "Unintended Consequences? A Look at Potential Farmer Impact." In *Facing Hawai'i's Future: Harvesting Essential Information about GMOs,* edited by Ana Currie, 39–43. Hilo, HI: Hawai'i SEED.

Bersani, Leo. 1987. "Is the Rectum a Grave?" *October* 43:197–222.

Bondera, Melanie. 2006. "Papaya and Coffee: GMO 'Solutions' Spell Market Disaster." In *Facing Hawai'i's Future: Harvesting Essential Information about GMOs,* edited by Ana Currie, 44–46. Hilo, HI: Hawai'i SEED.

Bondera, Melanie, and Mark Query. 2006. "Hawaiian Papaya: GMO Contaminated." http://www.grain.org/research_files/Contamination_Papaya.pdf.

Cline Harry. 2010. "GM Papaya Wins Approval in U.S., Japan." *Western Farm Press.* April 21. http://westernfarmpress.com/government/gm-papaya-wins-approval-us-japan.

Ferreira, Stephen, Ronald Mau, Karen Pitz, Richard Manshardt, and Dennis Gonsalves. 1993. "Papaya Ringspot Virus Cross Protection—An Update." In *Proceedings: 29th Annual Hawai'i Papaya Industry Association Conference, September 24–25, 1993,* edited by Chian Leng Chia and Dale O. Evans, 16–17. Honolulu: UHM

CTAHR (University of Hawai'i at Mānoa, College of Tropical Agriculture and Human Resources).

Foucault, Michel. 1990 [1976]. *The History of Sexuality, Volume 1.* Translated by Robert Hurley. New York: Vintage.

Franklin, Sarah. 2003. "Ethical Biocapital: New Strategies of Cell Culture." In *Remaking Life and Death: Toward an Anthropology of the Biosciences,* edited by Sarah Franklin and Margaret Lock, 97–128. Santa Fe, NM: School of American Research Press.

Gonsalves, Carol. 2001. "Transgenic Virus-Resistant Papaya: Farmer Adoption and Impact in the Puna Area of Hawai'i." MA thesis, Empire State College at State University of New York.

Gonsalves, Carol, David Lee, and Dennis Gonsalves. 2004. "Transgenic Virus-Resistant Papaya: The Hawaiian 'Rainbow' Was Rapidly Adopted by Farmers and Is of Major Importance in Hawai'i Today." *APSnet Features.* http://www.apsnet.org/publications/apsnetfeatures/Pages/PapayaHawaiianRainbow.aspx.

Gonsalves, Dennis. 1998. "Control of Papaya Ringspot Virus in Papaya: A Case Study." *Annual Review of Phytopathology* 36:415–37.

———. 2003. "The Papaya Story: A Special Case or a Generic Approach?" Paper presented at the annual meeting of the National Agricultural Biotechnology Council (NABC), Seattle, Washington, June 1–3.

Gonsalves, Dennis, and Stephen Ferreira. 2003. "Transgenic Papaya: A Case for Managing Risks of Papaya Ringspot Virus in Hawai'i." *Plant Health Progress.* http://www.plantmanagementnetwork.org/pub/php/review/2003/papaya/.

Gonsalves, Dennis, Carol Gonsalves, Steve Ferreira, Karen Pitz, Maureen Fitch, Richard Manshardt, and Jerry Slightom. 2004. "Transgenic Virus Resistant Papaya: From Hope to Reality for Controlling Papaya Ringspot Virus in Hawai'i." *APSnet Features.* http://www.apsnet.org/publications/apsnetfeatures/Pages/papayaringspot.aspx.

Gottlieb, Alan, and Lorie Farrell. 2011. "Vandalism at Isle Farms and Ranches Is about More than Just Property Damage." *Honolulu Star-Advertiser.* July 24.

Greenpeace. 2003. "Patented Papaya: Extending Corporate Control over Food and Fields." http://www.greenpeace.org/seasia/ph/Global/seasia/report/2003/5/patented-papaya.pdf.

———. 2006. "Papaya: The Failure of GE Papaya in Hawai'i." http://www.greenpeace.org/seasia/ph/Global/seasia/report/2006/6/copy-of-papaya-the-failure-o.pdf.

Hao, Sean. 2010. "Genetically Modified Papaya Problematic." *Honolulu Star-Advertiser.* April 25.

Haraway, Donna J. 1991. "Situated Knowledges: The Science Question in Feminism and the Privilege of Partial Perspective." *Feminist Studies* 14 (3): 575–599.

———. 1997. *Modest_Witness@Second_Millennium.FemaleMan©_Meets_OncoMouse™: Feminism and Technoscience.* New York: Routledge.

———. 2008. *When Species Meet.* Minneapolis: University of Minnesota Press.

Hawai'i Department of Agriculture. 2008. *Food Self-Sufficiency in Hawai'i.* http://Hawai'i.gov/hdoa/add/White%20Paper%20D14.pdf.

Hawai'i Papaya Industry Association. 2009. "The Rainbow Papaya Story." http://www.hawaiipapaya.com/rainbow.html.

Hayden, Cori. 2003. *When Nature Goes Public: The Making and Unmaking of Bioprospecting in Mexico.* Princeton, NJ: Princeton University Press.

Helmreich, Stefan. 2009. *Alien Ocean: Anthropological Voyages in Microbial Seas.* Berkeley: University of California Press.

Hine, Richard, Oliver Holtzmann, and Robert Raabe. 1965. *Diseases of Papaya (*Carica Papaya *L.) in Hawai'i.* Honolulu: UHM CTAHR. http://www.ctahr.Hawaii.edu/oc/freepubs/pdf/B-136.pdf.

Honolulu Star-Advertiser. 2011a. "Papaya Vandals Must Be Stopped." July 21.

———. 2011b. "Vandals Chop Down 10 Acres of Puna Papaya Trees." July 19.

Isherwood, Myron. 1994. "Regulations Governing Papaya Ringspot Virus Control." Honolulu: UHM CTAHR. http://www.ctahr.hawaii.edu/oc/freepubs/pdf/HITAHR_05-08-94_6-15.pdf.

Jinbo, Paige. 2011. "Vandalism Cases at Papaya Farm Turn into Dead End for Police." *Honolulu Star-Advertiser.* July 28.

Kakesako, Gregg. 2010a. "Vandals Chop Big Isle Papaya Trees." *Honolulu Star-Advertiser.* July 2.

———. 2010b. "Big Isle Farmer Now Tallies Papaya Loss at 13,000 Trees." *Honolulu Star-Advertiser.* July 17.

Kobayashi, Wayne. 1993. "Update on the Papaya Ringspot Virus Situation in Puna." In *Proceedings: 29th Annual Hawai'i Papaya Industry Association Conference, September 24–25, 1993,* edited by Chian Leng Chia and Dale O. Evans, 3–5. Honolulu: UHM CTAHR.

Kubota, Gary. 2011. "Decapitation of Papaya Trees Unnerves Hawai'i Isle Farmers." *Honolulu Star-Advertiser.* July 20.

Manshardt, Richard. 1993. "Update on Genetically Engineered PRV Resistance." In *Proceedings: 29th Annual Hawai'i Papaya Industry Association Conference, September 24–25, 1993,* edited by Chian Leng Chia and Dale O. Evans, 18–19. Honolulu: UHM CTAHR.

———. 2002. *Is Organic Papaya Production in Hawai'i Threatened by Cross-Pollination with Genetically Engineered Varieties?* Honolulu: UHM CTAHR. http://www.ctahr.Hawaii.edu/oc/freepubs/pdf/BIO-1.pdf.

National Agricultural Statistics Service (NASS). 2009. "Hawai'i Papayas." http://www.nass.usda.gov/Statistics_by_State/Hawaii/Publications/Fruits_and_Nuts/papaya.pdf.

———. 2010. "2009 Papaya Utilization Down." http://www.nass.usda.gov/Statistics_by_State/Hawaii/Publications/Fruits_and_Nuts/anpapFF.pdf.

Nishina, Melvin, Stephen Ferreira, Richard Manshardt, Catherine Cavalctto, Emerson Llantero, Loren Mochida, and Delan Perry. 1998. *Production Requirements of the Transgenic Papayas 'UH Rainbow' and 'UH SunUP.'* Honolulu: UHM CTAHR. http://www.ctahr.Hawaii.edu/oc/freepubs/pdf/NPH-2.pdf.

Nishina, Melvin, Wayne Nishijima, Francis Zee, Chian Leng Chia, Ronald Mau, and Dale Evans. 1989. *Papaya Ringspot Virus (PRV): A Serious Disease of Papaya.*

Honolulu: UHM CTAHR. http://www.ctahr.Hawaii.edu/oc/freepubs/pdf/CFS-PA-4A.pdf.

Parcon, Hazel, Run Yu, Matthew Loke, and PingSun Leung. 2010. *Competitiveness of Hawai'i's Agricultural Products in Japan.* Honolulu: UHM CTAHR. http://www.ctahr.Hawaii.edu/oc/freepubs/pdf/EI-19.pdf.

Pechar, Emily, and Toby Tatum. 2008. "Biotechnology: State-Level Legislative Analysis." http://www.opar.gtri.gatech.edu/report/report-download.do?reportId=7.

Pesante, Amy. 2003. "Market Outlook Report: Fresh Papayas." http://Hawaii.gov/hdoa/add/research-and-outlook-reports/papaya%20outlook%20report. pdf.

Pew Initiative on Food and Biotechnology. 2005. "State Legislative and Local Activities Related to Agricultural Biotechnology Continue to Grow in 2003–2004." http://www.pewtrusts.org/uploadedFiles/wwwpewtrustsorg/Fact_Sheets/Food_and_Biotechnology/PIFB_State_Legislature_2003–2004Session.pdf.

———. 2007. "State Legislative Activity Related to Agricultural Biotechnology in 2005–2006." http://www.pewtrusts.org/uploadedFiles/wwwpewtrustsorg/Reports/Food_and_Biotechnology/PIFB_State_Legislature_2005–2006Session.pdf.

Pope, Willis. 1930. *Papaya Culture in Hawai'i.* Bulletin (Hawai'i Agricultural Experimental Station) Number 61. Washington, DC: Government Printing Office.

Salazar, Stephanie, and Baron Sekiya. 2011. "Video: Hawai'i Papaya Farms Attacked, Farmers Hui in Puna." http://www.bigislandvideonews.com/2011/07/20/video-Hawaii-papaya-farms-attacked-farmers-hui-in-puna/.

Storey, W. B. 1941. "The Botany and Sex Relationships of the Papaya." In *Papaya Production in the Hawaiian Islands,* 5–22. Bulletin (Hawai'i Agricultural Experimental Station) Number 87. Honolulu: Hawai'i Agricultural Experiment Station.

Tripathi, Savarni, Jon Suzuki, and Dennis Gonsalves. 2006. "Development of Genetically Engineered Resistant Papaya for Papaya Ringspot Virus in a Timely Manner: A Comprehensive and Successful Approach." In *Plant-Pathogen Interactions: Methods and Protocols,* edited by Pamela Ronald, 197–239. Totowa, NJ: Humana Press.

Weasel, Lisa. 2009. *Food Fray: Inside the Controversy over Genetically Modified Food.* New York: AMACOM.

Yee, W., G. M. Aoki, R. A. Hamilton, F. H. Haramoto, R. B. Hines, O. V. Holtzmann, J. T. Ishida, J. T. Keeler, and H. Y. Nakasone. 1970. *Papayas in Hawai'i.* Circular (University of Hawai'i, Cooperative Extension Service), 436. Honolulu: University of Hawai'i Cooperative Extension Service.

Yeh, Shyi-Dong, Huey-Jiunn Bau, Yi-Jung Kung, and Tsong-An Yu. 2007. "Papaya." *Biotechnology in Agriculture and Forestry* 60:73–96.

Yoon, Carol Kaesuk. 1999. "Stalked by Deadly Virus, Papaya Lives to Breed Again." *New York Times.* July 20. http://www.nytimes.com/1999/07/20/science/stalked-by-deadly-virus-papaya-lives-to-breed-again.html?scp=1&sq=stalked+by+deadly+virus+yoon&st=nyt.

6 | Seeds of Contestation

The Emergence of Hawai'i's Seed Corn Industry

Benjamin Schrager and
Krisnawati Suryanata

> I am not going to sit here and lament sugar and lament pineapple gone. That was a monoculture that had its time, had its place in other centuries. That's not the 21st century. Many of the people in this room today represent the seed industry, which is I think the new foundation and basis for agricultural prosperity in the state.
>
> Neil Abercrombie, then governor of Hawai'i, addressing the Hawai'i Crop Improvement Association Annual Meeting (May 16, 2011)

In spite of its reduced contribution to the state's economy, agriculture continues to be of special concern in Hawai'i. Agriculture occupies a large land area and is closely associated with people's cultural heritage and attitudes. Since Hawai'i became a state in 1959, a number of legislative actions and policies have been adopted to support agricultural preservation (Suryanata and Lowry, this volume). This goal resonates with the state's legacy of a strong agrarian economy as well as holistic Native Hawaiian agro-food systems (Kame'eleihiwa, this volume). When the plantation economy declined precipitously in the 1990s, many state and community leaders rallied to push for diversified agriculture, which includes all agricultural industries other than sugar and pineapple. Unfortunately, efforts to diversify agriculture in the past have stumbled due to global competition (Suryanata 2000, 2002).

Since the early 2000s, however, the size of the seed corn industry in Hawai'i rapidly increased. Some of the largest seed companies, including Monsanto, Pioneer Dupont, Syngenta, Dow AgroSciences, and BASF, have established themselves in the state. In terms of economic investments in staff and infrastructure, the seed corn industry in Hawai'i represents the largest concentration of tropical seed corn nurseries in the world. Thus, for the first time since the mid-twentieth century, one particular sector enjoyed extremely robust growth. Many see the seed corn industry as representing a promising future for agriculture in Hawai'i, a view reflected in then Governor Abercrombie's statement cited above. But others saw the perils of allowing global agribusinesses to dominate the state's agricultural landscape. The fact that the seed corn companies utilize genetic engineering and produce genetically modified organisms (GMOs) causes further concerns.

As the contentious debate grew, county legislators in Kaua'i, Hawai'i, and Māui proposed bills aiming to curb the growth of the seed corn industry. Each of these proposals involves lengthy battles that have divided the islands' communities like never before. Bill proponents rally around opposition to GMOs and pesticides, claiming that seed corporations are conducting dangerous experiments that adversely impact human health and the environment. Signs proclaiming "no GMOs" that accuse the seed industry of "poisoning Hawai'i" are ubiquitous along roadsides, at political gatherings, and even in surfing competitions. Although the resulting ordinances were later struck down in federal court, this legislative success galvanizes the anti-GMO coalition because they see these legislative successes at the county level as a promising step towards food democracy.

We contend that the current debate is typically framed in terms of binaries such as GM/organic, good science/bad science, and global/local that fail to illuminate the underlying processes that made the industry so influential in the first place. The issues surrounding the rise of the seed corn industry are complex, encompassing political economic and technological developments that are currently outside the scope of the public debate. For instance, we do not know if the seed corn industry can continue to provide stable employment and economic opportunities in Hawai'i as the governor expressed in his speech. Likewise, preventing seed companies from using GM crops in their operations in Hawai'i would address neither the extant problems that face small farmers in Hawai'i, nor consumers' concerns about GMO contamination in their food. More importantly, neither position informs the citizenry about the true challenges that they face in building an alternative food system.

Our analysis situates Hawai'i's seed corn industry within the development of industrial corn, the technological change in corn breeding, the organizational restructuring in the seed corn industry, and the changes in Hawai'i's agrarian landscape. We discuss how advances in genetic engineering and molecular breeding allow the seed corn companies to deterritorialize their operations as they simultaneously speed up the rate of crop improvement. We apply this analysis to the case of the seed corn industry in Hawai'i and argue that a convergence of forces in the 2000s led to Hawai'i being prominently positioned just as seed corporations sought to expand their year-round seed corn nurseries. We examine how the seed industry's rapid growth has been accompanied by increased opposition to seed corporations that are primarily articulated in technical terms. Finally, we evaluate the implications of the industry's flexible accumulation strategy on the future of the seed corn industry in Hawai'i.

Research Methodology

We develop a set of working questions to investigate the dramatic growth of the seed corn industry in Hawai'i. The first question reviews the development of the global industrial corn network, the technical changes within corn breeding, and the position of Hawai'i's seed corn industry within this evolving network. The second question situates the seed corn industry within Hawai'i's agricultural landscape, and examines the ways that land use and agricultural policies—as well as anti-GMO citizens' movements—affect the seed corn industry.

Information for this study came from interviews, public events, videos posted on the Internet, and archival materials that include academic literature on crop breeding, legislative documents, corporate reports, newspaper archives, and industry records—many of which came from the Hawai'i Crop Improvement Association (HCIA), a group that promotes the seed industry. The lead author met with seventeen different informants including industry representatives, corporate executives and staff, academic researchers, and community leaders. The semistructured interviews ranged from casual conversations of fifteen minutes in length to recorded interviews of several hours. We sought different information based on the expertise of the interviewee. For example, when speaking with industry proponents, the interviews would focus on the logistics, science, and economics of the seed industry. When speaking with the industry's critics, we sought to understand the range of their concerns. When speaking with government employees, we discussed the evolution of regulatory institutions and policies.

Gaining access to industry actors proved challenging. Seed corporations are notoriously secretive and suspicious of researchers. To overcome these barriers, the lead author met plant scientists at the University of Hawai'i, who then made the introduction to key personnel in the seed industry. Throughout the project we strove to develop collaborative as opposed to extractive relationships with our interviewees (Lassiter 2005), be sensitive to the dynamics that arise from interviewing people with power (Rice 2010), and share the outcome of the research with our interviewees. We attended fourteen public events related to agriculture and the seed corn industry in Hawai'i in 2012–2013. The format of these public events included panel discussions, public hearings, workshops, conferences, festivals, and protests. Attending these events helped us understand the unique set of issues associated with Hawai'i's seed corn industry.

Industrialization of Seed Corn Production

Modern industrial agriculture is characterized by vertically integrated specialized agricultural operations, in which the seed industry is a key component. A single farmer or corporate entity must coordinate with other operations that produce vast quantities (Fitzgerald 2003; Hart 2003). The increased productivity led to burgeoning surpluses and low grain prices, which in turn were key to the intensive regime of accumulation that allowed the post-war growth of US economy. To support US farmers, the government subsidizes grain producers in the form of price supports (Dixon and Hapke 2003). Because of these price supports, US grain farmers can disregard the market signals generated by overproduction and continue on maximizing their yields in order to increase their revenues, a strategy referred to as "productionism" (Lang and Heasman 2004; Thompson 1995).

Corn is the number one commodity that is produced under the productionist strategy (Goodman et al. 1987; Warman 2003). Breeders seek to develop varieties of corn that produce higher yields by manipulating ear location, kernel distribution, ability to remain upright, pest resistance, and disease resistance. A major innovation that changed corn breeding and corn industry is the development of hybrid corn in the early twentieth century (Fitzgerald 1990; Kloppenburg 2005). To develop hybrid corn, corn is self-pollinated (a.k.a. selfing) for several generations before the inbred cultivar is considered genetically homogenous and stable. Using a technique known as backcrossing, corn breeders can incorporate desired traits into a higher quality inbred parent line. Corn breeders then cross-pollinate distinct parent lines to create robust "hybrid" varieties that are tested for commercial viability. Developing parent lines through selfing or backcrossing is a temporally lengthy process, because it takes six to eight generations to complete.

The next generation of seeds produced from hybrid corn plants lose their uniformity as well as their hybrid vigor through genetic segregation, hence hybrid varieties cannot be replanted without significant losses in yield. This genetic property gave seed producers greater control over the benefits of seed improvement, thus providing incentives for private firms to invest in seed corn production. Kloppenburg (2005, 93) writes, "[Hybridization] . . . uncouples seed as 'seed' from seed as 'grain' and thereby facilitates the transformation of seed from a use-value to an exchange-value." Nevertheless, while farmers had to purchase hybrid seeds, they quickly preferred to grow hybrid corn to open-pollinated corn because the nominal cost charged by land-grant universities belied the value hybrid corn generated for farmers (Bogue 1983).

In Iowa, hybrid corn production increased from under 10 percent in 1935 to over 90 percent of corn production in 1940 (Bogue 1983). After Iowa farmers adopted hybrid corn, it spread gradually out from Iowa to the rest of the Corn Belt and then beyond (Griliches 1960). By 1960, nearly all corn acreage in the United States was planted to hybrids (Hallauer et al. 2010). The increased production results in a chronic over-capacity of corn production, along with the proliferation of innovations in downstream uses of corn such as livestock feed or fructose industry, which utilize the surplus grain to substitute for interchangeable inputs (Goodman et al. 1987). Of all recent industrial uses for corn, the largest increase is in the demand for ethanol, which is a biofuel primarily used in automobiles. The 2005 and 2007 energy bills along with the 2008 farm bill further encouraged dramatic increases in corn-based ethanol production (Lehrer 2010) to build what McMichael (2009) calls the "food-fuel complex." Today an immense amount of capital flows through industrial corn networks, and the seed corn production is a large and lucrative industry.

Technoscientific Change and Corporate Restructuring

As early as the 1940s, emerging seed companies worked alongside land-grant institutions in the development and marketing of commercial corn lines. Over time, seed companies gradually displaced land-grant universities as the purveyors of hybrid seeds (Fitzgerald 1990; Goodman et al. 1987; Kloppenburg 2005; Lacy and Busch 1989). As breeding research became more capital intensive, a few seed companies were better situated to make research and development investments in staff, infrastructure, and distribution networks. Seed corn companies also used mergers and acquisitions to acquire proprietary seed stock from their competitors.

From the 1980s, public sector research for crop improvement shifted towards the emerging field of agricultural biotechnology that combined molecular biology and crop improvement (Knight 2003). Researchers at land-grant universities "sought access to prestigious federal grants (e.g., NSF, NIH) that supported molecular biology research" (Buttel 2005, 279), which left crop breeding in the hands of private seed companies. In addition, legal developments in establishing rights to new plant varieties, founded upon the 1930 Plant Patent Act and the 1970 Plant Variety Act, created incentives for innovation and capitalization in breeding (Haraway 1997). The 1980 *Diamond v. Chakrabarty* US Supreme Court decision was a landmark case that allows for the patenting of deoxyribonucleic acid (DNA) sequences or organisms into which those DNA sequences were inserted. Haraway (1997) describes how these rulings on biotechnology problematically extend private ownership over living organisms.

One corporation in particular, an agricultural chemical corporation from St. Louis called Monsanto, began aggressively pursuing agricultural biotechnology, including genetic engineering (GE). Monsanto is a useful case study, because in the mid-1990s Monsanto was the first corporation to develop commercially viable genetically modified (GM) crops (Schurman and Munro 2010, 18). The two leading GM crops were Round-up Ready soybeans that are engineered to be resistant to Round Up (glyphosate) herbicide, and Bt (*bacillus thuringiensis*) corn that is engineered to produce Bt toxin that is lethal to pests such as corn borer. For both of these GM traits, Monsanto had successfully developed and patented a DNA sequence, but did not produce commercially viable GM seeds on their own.

In the late 1990s, Monsanto revolutionized the seed crop industry by expending $6.5 billion in separate acquisitions to purchase Holden Seeds, DeKalb, Delta and Pine Land, and Cargill International so they could integrate the GM traits into their own proprietary germplasm (Schurman and Munro 2010; Charles 2001). Monsanto was the first entity to integrate traditional crop improvement, biotech research, and agrochemical research. All of Monsanto's competitors were forced to adapt to this new corporate structure to stay competitive. When the dust had settled, a handful of seed corporations—Monsanto, DuPont Pioneer, Syngenta, Dow AgroSciences, Bayer, and BASF—emerged to dominate the supply of seeds, agrochemicals, and GM traits. The top four companies now control approximately 80 percent of seed corn marketed in the United States (Kronberg 2012), and GM corn varieties are now more common than conventional corn varieties in the United States. The area planted to GM corn grew from 25 percent in 2000 to 80 percent in 2008 and 90 percent in 2013 (Fernandez-Cornejo et al. 2014).

Crop breeding underwent another revolutionary change with the application of advances in molecular biology, and corn is at the forefront of this development. Traditionally, breeders rely on phenotype—traits that are physically observable—to develop improved varieties. Trait expression, however, is affected by variable environmental conditions or the development stage of the plant. In the 1980s, advances in DNA or molecular markers allowed scientists to develop new techniques that link key genetic sequences with DNA markers, and use them for marker-assisted selection (MAS). Using this technique, breeders can use the presence or absence of a marker as a substitute for phenotypic selection (Collard et al. 2005; Moose and Mumm 2008; Xu et al. 2012). MAS increases breeders' precision in selecting superior cultivar genotypes because they are not misled by the environmental component of phenotypic observations. Additionally, selection can be carried out at the seedling or even seed stage (Sakiyama et al. 2014), which is especially advantageous for traits that are expressed at later developmental stages. By identifying inferior genotypes early, MAS helps breeders reduce the size of the effective population they work with. MAS thus makes the selection process faster, more efficient, and more reliable (Collard and Mackill 2008; Collard et al. 2005). By using MAS, complex field trials that need to be conducted at particular times of year or at specific locations can be substituted with molecular tests, which allow selection in off-season nurseries that can grow more generations per year (Moose and Mumm 2008; Collard and Mackill 2008). Indeed, year-round nurseries and MAS augment each other and are now central to seed corporations' crop improvement strategies—propelling places such as Hawai'i to be an important site in corn seed production.

Hawai'i's Seed Corn Industry (HSCI)

Corn breeders began to use Hawai'i as a winter nursery since the late 1960s, when they brought their most promising cultivars to Hawai'i. As we discussed earlier, conventional breeding of hybrid corn requires six or seven generations of self-pollination and backcrossing to produce parent lines. In tropical locations such as Hawai'i, crop breeders could produce one extra generation during the winter to shorten the time required for seed corn improvement. Cultivars were typically planted, self-pollinated, and then shipped back to the mainland for further trials and selections (Brewbaker 2003). In 1968, Professor James Brewbaker from the University of Hawai'i at Mānoa founded the Hawai'i Crop Improvement Association (HCIA), which later became instrumental to the development of seed corn nurseries in Hawai'i. According to its founding constitution, HCIA's original purposes were to help crop breeders address the agronomic challenges posed by Hawai'i's environment and to help facilitate a seed certification service. HCIA sponsored an annual conference and the attendees included leaders from the seed corn industry such as staff from the agricultural college, employees of sugar and pineapple plantations, politicians, and mainland as well as Hawai'i crop breeders. Not only were winter nurseries beneficial for crop improvement, a trip to Hawai'i was also a welcome reprieve for crop breeders arriving from the subfreezing Midwest. For instance, in 1971 an Indiana agricultural company offered a special vacation package called "Seedmen's Hawaiian Holiday" that included "jet airfare, resort hotel accommodations on Waikiki Beach, and a day at Molikai [*sic*] Seed Service."

Corn research at the University of Hawai'i examined the industry's potential as a part of the agricultural diversification strategy (Brewbaker 2003). Throughout the 1980s, researchers at the University of Hawai'i investigated whether farmers in Hawai'i could grow commodity corn (e.g., Chung et al. 1982; Thomassin et al. 1985). But Hawai'i's high costs of production renders commodity corn production not competitive when compared to the Corn Belt (Singh 1983), leaving only a niche production of sweet corn hybrids (Brewbaker 1982). Seed corn production, however, was considered to be more promising because of the locational advantage of operating winter nurseries.

Throughout the 1980s and 1990s, sugar and pineapple industries were declining rapidly, displacing workers and releasing large tracts of prime agricultural lands. State and industry leaders sought opportunities to diversify the agricultural sector and hoped to expand the seed corn industry. For example, a task force in 1983 commissioned by the then mayor of Māui County examined the consequences of the closure of pineapple operations in Molokai, and recommended expanding the seed corn operations on the island (Bowen and Foster 1983). Seed companies that previously had difficulty gaining access to land saw an opening that would allow them to expand their operations. For example, a former sugar plantation manager explained how a representative from a seed company approached him after an accidental fire burned up about forty acres of sugar land. He described this transaction in an interview:

> [The seed company employee] said, "Do you want to sublease the land for me to grow seed corn?" So I took him up there and his eyes got as big as saucers. "Oh. I can have that?" . . . I told him that he could also have the water from the ditch for free. . . . He was absolutely thrilled, and I was absolutely thrilled because I was making a lot more money leasing it to him for seed corn than I was for growing sugar. (Interview with first author on January 23, 2013)

When Hurricane Andrew in 1992 badly damaged the seed corn industry in southern Florida, some leaders within HCIA pointed out the opportunity for Hawai'i to fill the void. In a letter announcing the 1993 HCIA meeting, then HCIA President Robert Osgood wrote, "Most of us would like to see a much larger seed industry expanding beyond the predominant winter seed nursery operations." In spite of these efforts, the seed corn industry would remain a minor industry in Hawai'i until the late 1990s when its value began to skyrocket (figure 6.1).

Several forces converged, which led the seed companies to rapidly increase their investment in Hawai'i. The first set of forces are related to broader changes in the political economy of industrial corn production

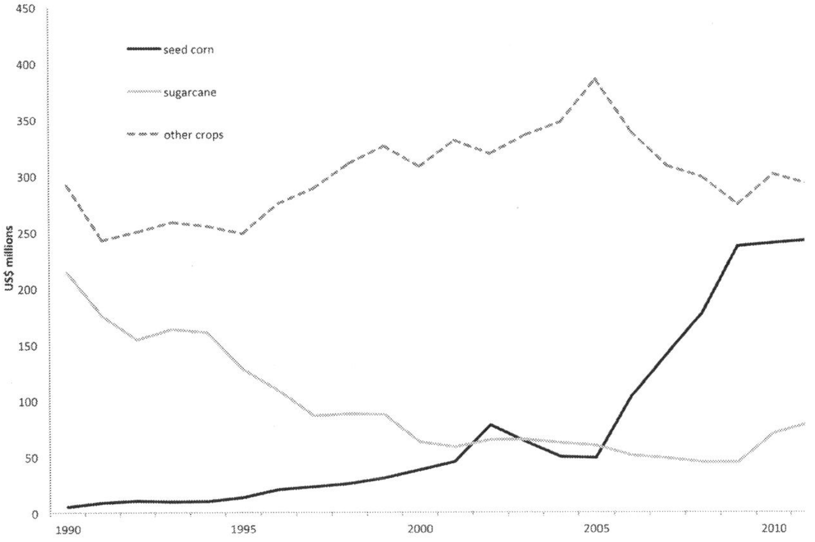

FIGURE 6.1. Value of Hawai'i's crop sales (unadjusted), 1990–2011. *Source*: USDA, National Agricultural Statistics Service.

that was fueled by the ethanol boom and the dramatic growth of commercial GM corn. While seed corn production had been a large and profitable industry for decades, it has now become even more lucrative. After the corporate consolidation in the late 1990s, only a few major seed companies were left in intense competition with each other. They heavily increased their capital investment in order to remain competitive and reap as much profits from the burgeoning corn seed industry as possible.

The second set of forces concern the technoscientific changes in crop improvement discussed previously. The development of GM traits and advances in MAS provide tools for seed companies to increase the pace of crop improvement in corn. As breeders rely less on location-specific phenotypic data for cultivar selection, operating ex situ year-round nurseries became a key strategy for corn seed companies to get the latest genetic information into the pipeline faster, hence increasing the pace of crop improvement. These two broader changes induced seed corn corporations to increase their capital investments, and Hawai'i emerged as an ideal location to be incorporated in seed corporations' accumulation strategy.

Hawai'i's Comparative Advantages

Many would point out that Hawai'i's primary advantage lies in its climate that allows corn to grow year-round. For example, an industry insider observes: "Mexico has rainy season, Puerto Rico has hurricanes, and Chile and Argentina have a winter." Another concurs: "Hawai'i is very reliable. . . . very seldom that we get a crop failure here" (Paiva 1999). But beyond its favorable climate, three other factors converged in the late 1990s that made Hawai'i an ideal location for the seed corporations to quickly expand their investment in year-round nurseries.

First, the Land Use Law and farm preservation policies have preserved large tracts of prime agricultural lands throughout Hawai'i, in spite of the general decline of agriculture over the past half century (Suryanata and Lowry, this volume). Developers wishing to convert land out of agricultural classification must navigate a lengthy process that involves assessing the environmental impacts, allowing for citizens' input, and getting approval from the Land Use Commission. As a result, prime agricultural land, much of which has valuable infrastructure such as irrigation and road network, was amply available when the seed corporations were poised to expand their operations in the early 2000s.

To illustrate, some of the most desirable agricultural land in Hawai'i is located on the western or leeward sides of Kaua'i, Māui, and O'ahu. These sides have less cloud cover and therefore receive more solar radiation. At the start of the twentieth century, sugar plantations built irrigation systems that diverted water from the rainier windward side to irrigate crop production on the leeward side, hence benefitting from both water and sunshine. Seed corporations were able to lease or purchase these lands to cultivate seed corn. Today HSCI is estimated to occupy approximately 25,000 acres (Callis 2013) throughout the state, although they only cultivate about a fourth of their acreage annually due to various pollination and pest management strategies.

Second, Hawai'i has a strong labor force with a combination of skilled and unskilled workers to carry out the three main tasks in operating year-round nurseries: cultivar selection, seed increase, and observation. While cultivar selection and observation are more specialized tasks, seed increase is straightforward crop propagation to ramp up production. Most of the skilled workers are university graduates trained in agricultural sciences, from both the continental United States as well as from Hawai'i.

Seed corn operations also employ a large number of unskilled workers to conduct agronomic and logistical tasks in seed corn improvement, in which hand pollination is the most basic and time-consuming task. It involves taking the pollen from a corn tassel and placing it in a bag over the silks of the appropriate plant. These jobs are repetitive and the worker must carefully follow the instructions so that the correct pollinations are carried out. Many of these workers are recent Filipino immigrants and signs on seed companies' operations are often posted in both English and Filipino languages. The seed companies directly manage seed corn nurseries as well as contract them to local farms, which allow the companies to adapt to seasonal and fluctuating labor needs.

Finally, the College of Tropical Agriculture and Human Resources of the University of Hawai'i and the HCIA are key organizations that facilitated the rapid growth of seed corn nurseries in Hawai'i. There are numerous connections between HSCI and scientists working at the agricultural college. For example, Monsanto donated $500,000 in 2011 to create the Monsanto Scholarship Fund at the agricultural college (Schrire 2011), and seed corporations are the largest employer of agricultural scientists with advanced degrees in the state. HCIA has evolved to become an industry trade group that helps seed corn companies to navigate the agronomic and political barriers in the state. Thus, the institutional networks were in place when year-round nurseries became more central to seed companies' corporate strategy.

Industry representatives insist that the increased investment in year-round nurseries in Hawai'i is a business decision because they have consistently received good return for their investment. The most valuable contribution of Hawai'i's seed corn industry is in speeding up the research process in cultivar development. In this way, companies that invest in year-round operations gain a significant advantage over those that do not. As technological and organizational changes reconfigured seed corporations in the late 1990s, Hawai'i was well positioned as the logical place for companies to quickly expand their year-round seed corn nurseries.

The seed corn industry is now a dominant sector in the agricultural economy of Hawai'i (see figure 6.1). The industry's value grew from $5.1 million in 1990 to $241.6 million in 2011, increasing its share from less than

1 percent to 40 percent of Hawai'i's crop revenues. Meanwhile, sugar and pineapple's combined value decreased by more than half (*Statistics of Hawai'i Agriculture* multiple years). In a study commissioned by HCIA, Loudat and Kasturi (2013) claim that in 2012 HSCI directly and indirectly contributed $551 million to Hawai'i's economy. Furthermore, they point out that in addition to generating employments and tax revenue within the state, the seed corn industry keeps prime agricultural land in agricultural production and therefore helps fulfill the goal of preserving agriculture in the state.

Growing Discontent

The fact that the seed companies have quietly grown to be the new "Big Five" in a relatively short period of time caught many Hawai'i residents by surprise. Furthermore, the dramatic growth of the seed corn industry throughout the 2000s is inextricably linked to the development of GM crops, which remains the world's most vigorously contested agricultural technology. Organizational and technological changes in the seed corn industry have also made the operation more opaque and difficult for the general public to understand. While there are visible fields of seed corn, the manners and purposes of their operations remain mysterious to most casual observers.

Hawai'i residents have a long history of active involvement in projects and industries that affect the local environment and community. Citizen actions have effectively derailed a number of high-profile projects such as the Hokulia housing development project on the Island of Hawai'i (Carlton 2005; Suarez 2004) and the Super Ferry (Paik and Mander 2009). In each of these controversies, distrust against dominant (outside) capital which is perceived to not care about local environment and community prevails, especially in rural Hawai'i. The (in)visibility of seed corporations in Hawai'i caused the seed corn industry to become a setting where anti-GMO activists can raise numerous doubts to mobilize a coalition opposing the seed corn industry. For example, activists have accused the companies of applying pesticides excessively and running GM trials in open fields (Freese et al. 2015). Furthermore, the seed companies' policy is to keep their field locations confidential. While threats posed by vandalism, theft, and physical harm are important to consider, the companies' refusal to disclose the location of their operations further reinforces the barrier between the new industry and the general public. In doing so, it inadvertently raises their vulnerability to speculations that seed corporations are furtively taking over Hawai'i's prime agricultural land, or worse, have something to hide in their operations. As the controversy garnered momentum, news media have suggested that Hawai'i has become "ground zero for GMOs" (Harmon 2014; McAvoy 2014; Mitra 2014; Pollack 2013).

It is noteworthy that unlike the anti-GMO movements on the mainland that focus on labeling to protect consumers, the anti-GMO movement in Hawai'i has focused on GMO crop production, especially those operated by large seed companies. In response, the seed corporations use legal avenues to thwart legislation or lawsuits that target their operations. Support and funding from outside the state have poured in to contest seed companies' operations (DePledge 2013). A protracted legal battle is costly for both sides, but the stakes are also high. Seed corporations do not want to set a precedent that local governments can impose additional layers of scrutiny above and beyond federal regulations. Anti-GMO activists, on the other hand, view the legislative victories at the county level, even if they eventually do not stand the legal challenges at the state or national level, as critical to building momentum in the fight against the expansive growth of the GM industry.

While Hawai'i in the late 1990s might seem to be the most ideal site for seed companies to invest, it is not the only region suitable for seed corn nurseries to operate. A change of the favorable geographical conditions that had attracted the seed companies to invest in Hawai'i, relative to other regions, could change the investment climate in a relatively short period of time. The brewing anti-GMO controversy has no doubt tarnished the reputation of Hawai'i as the ideal place to do business. Meanwhile, heightened competition between regions and nations encourages the production of places with favorable business climates and special qualities for operating year-round nurseries. While Kaua'i County and Māui County recently sought to tighten regulations on seed corporations, in 2009 Puerto Rico passed a law titled "Law for the Promotion and Development of Agricultural Biotechnological Businesses in Puerto Rico."

Conclusion

The intense contestation surrounding seed corporations' operations in Hawai'i demonstrates the divergent visions with regard to the future of agriculture in Hawai'i. Industry proponents view the growth of seed corn industry as an opportunity for Hawai'i to salvage its agrarian economy. They tout the economic contributions of the industry such as agricultural employment, agricultural land preservation, and the development of agricultural infrastructure. Opponents, however, are unable to embrace an industry that is secretive and considered by many to embody what is undesirable about large agri-businesses. Indeed, many view the rise of the seed corn industry in Hawai'i to be an impediment to the development of local agro-food systems. As Martin (2005, 41) points out, in a polarized debate not only do the two sides become entrenched, but the terms of the debate themselves get entrenched. In Hawai'i, it has the effect of equating the rise of seed corporations with the excessive use of agrochemicals. Unfortunately, other important concerns such as economic alienation and corporate dominance over the landscape and foodscape become subordinated to the technical discussion of the levels of pesticide use and GM crop nurseries.

Our analysis shows the importance of examining the broader political economic environment, techno-scientific changes, and corporations' accumulation strategies that can lead to dramatic fluctuations as seed corporations swiftly shift, expand, and contract their off-season and year-round nurseries. This more nuanced understanding is critical in assessing the long-term position of Hawai'i in the global seed industry, and to effectively channel the engagement of those mobilized by the controversy towards food democracy.

References

Bogue, Allan G. 1983. "Changes in Mechanical and Plant Technology: The Corn Belt, 1910–1940." *The Journal of Economic History* 43 (1): 1–25.

Bowen, Richard L., and David L. Foster. 1983. "A Profile of Displaced Pineapple Workers on Moloka'i." University of Hawai'i at Mānoa, College of Tropical Agriculture and Human Resources, Research Extension Series, Vol. 31.

Boyd, William, W. Scott Prudham, and Rachel A. Schurman. 2001. "Industrial Dynamics and the Problem of Nature." *Society & Natural Resources* 14 (7): 555–570.

Brewbaker, James L. 1982. "Crop improvement in Hawai'i—Past, Present, and Future." University of Hawai'i at Mānoa, College of Tropical Agriculture and Human Resources, Vol. 180.

———. 2003. *Corn Production in the Tropics: The Hawaii Experience*. University of Hawai'i at Mānoa, College of Tropical Agriculture and Human Resources.

Buttel, Frederick H. 2005. "Ever since Hightower: The Politics of Agricultural Research Activism in the Molecular Age." *Agriculture and Human Values* 22 (3): 275–283.

Callis, Tom. 2013. "GMO: $243 Million Seed Industry Largest in State's Ag Sector." *West Oahu Today*, June 9. http://westhawaiitoday.com/sections/news/local-news/gmo243-million-seed-industry-largest-state%E2%80%99s-ag-sector.html.

Carlton, Jim. 2005. "Land-Use Ruling Shakes Hawaii Developers." *Wall Street Journal*. July 6. B.1.

Charles, Daniel. 2001. *Lords of the Harvest: Biotech, Big Money, and the Future of Food*. Cambridge: Perseus.

Chung, Jay H., James L. Brewbaker, and James L. Ritter. 1982. "Effects of Increasing Population Density on the Production of Corn in Hawaiʻi." University of Hawaiʻi at Mānoa, College of Tropical Agriculture and Human Resources, Vol. 13.

Collard, B. C. Y., M. Z. Z. Jahufer, J. B. Brouwer, and E. C. K. Pang. 2005. "An Introduction to Markers, Quantitative Trait Loci (QTL) Mapping and Marker-Assisted Selection for Crop Improvement: The Basic Concepts." *Euphytica* 1421 (1–2): 169–196.

Collard, Bertrand C. Y., and David J. Mackill. 2008. "Marker-Assisted Selection: An Approach for Precision Plant Breeding in the Twenty-First Century." *Philosophical Transactions of the Royal Society B-Biological Sciences* 363 (1491): 557–572.

DePledge, Derrick. 2013. "Mainland Cash Fuels the Fight." *Honolulu Star-Advertiser*, September 29. http://www.staradvertiser.com/hawaii-news/mainland-cash-fuels-the-fight/.

Dixon, Deborah P., and Holly M. Hapke. 2003. "Cultivating Discourse: The Social Construction of Agricultural Legislation." *Annals of the Association of American Geographers* 93 (1): 142–164.

Fernandez-Cornejo, Jorge, Seth Wechsler, Mike Livingston, and Lorraine Mitchell. 2014. "Genetically Engineered Crops in the United States." United States Department of Agriculture, Economic Research Service Report Number 162.

Fitzgerald, Deborah K. 1990. *The Business of Breeding: Hybrid Corn in Illinois, 1890–1940*. Ithaca, NY: Cornell University Press.

———. 2003. *Every Farm a Factory: The Industrial Ideal in American Agriculture*. New Haven, CT: Yale University Press.

Freese, Bill, Ashley Lukens, Alexis Anjomshoaa, and Sharon Perrone. 2015. *Pesticides in Paradise: Hawaii's Health and Environment at Risk*. Honolulu: Hawaii Center for Food Safety.

Goodman, David, Bernardo Sorj, and John Wilkinson. 1987. *From Farming to Biotechnology: A Theory of Agro-Industrial Development*. New York: Blackwell.

Griliches, Zvi. 1960. "Hybrid Corn and the Economics of Innovation." *Science* 132 (3422): 275–280.

Hallauer, Arnel R., Marcelo J. Carena, and Miranda J. B. Filho. 2010. *Quantitative Genetics in Maize Breeding*. New York and London: Springer.

Haraway, Donna J. 1997. *Modest_Witness@Second_Millennium.FemaleMan©_Meets_OncoMouse™: Feminism and Technoscience*. New York: Routledge.

Harmon, Amy. 2014. "A Lonely Quest for Facts on Genetically Modified Crops." *New York Times.* January 4. http://www.nytimes.com/2014/01/05/us/on-hawaii-a-lonely-quest-for-facts-about-gmos.html.

Hart, John F. 2003. *The Changing Scale of American Agriculture.* Charlottesville: University of Virginia Press.

Hofschneider, Anita. 2015. "Hawaii Is Feeling the Seed Industry's Downturn." *Civil Beat.* http://www.civilbeat.com/2015/06/hawaii-is-feeling-the-seed-industrys-downturn/.

Kloppenburg, Jack R. 2005. *First the Seed: The Political Economy of Plant Biotechnology.* Madison: University of Wisconsin Press.

Knight, Jonathan. 2003. "Crop Improvement: A Dying Breed." *Nature* 421 (6923): 568–570.

Koebner, Robert M. D., and Richard W. Summers. 2003. "21st Century Wheat Breeding: Plot Selection or Plate Detection?" *Trends in Biotechnology* 21 (2): 59–63.

Kronberg, Gene. 2012. "Overview of the Seed Industry." *AgriMarketing* 50 (8): 35–40.

Lacy, William B., and Lawrence Busch. 1989. "The Changing Division of Labor between the University and Industry: The Case of Agricultural Biotechnology." In *Biotechnology and the New Agricultural Revolution,* edited by J. J. Molnar and H. W. Kinnucan, 21–50. Boulder, CO: Westview Press.

Lang, Tim, and Michael Heasman. 2004. *Food Wars.* London and Sterling, VA: Earthscan.

Lassiter, Luke E. 2005. *The Chicago Guide to Collaborative Ethnography.* Chicago: University of Chicago Press.

Lehrer, Nadine. 2010. "(Bio)Fueling Farm Policy: The Biofuels Boom and the 2008 Farm Bill." *Journal of the Agriculture, Food, and Human Values* 27 (4): 427–444.

Loudat, Thomas, and Prahlad Kasturi. 2013. "Hawaii's Seed Crop Industry: Current and Potential Economic and Fiscal Contributions." Honolulu: Hawaii Farm Bureau and Hawaii Crop Improvement Association.

Mann, Susan A., and James M. Dickinson. 1978. "Obstacles to the Development of Capitalist Agriculture." *Journal of Peasant Studies* 5:466–481.

Martin, Brian. 2005. "Agricultural Antibiotics: Features of a Controversy." In *Controversies in Science and Technology: From Maize to Menopause,* edited by D. L. Kleinman, A. J. Kinchy, and J. Handelsman, 37–51. Madison: University of Wisconsin Press.

McAvoy, Audrey. 2014. "Why Hawaii Is Ground Zero for the GMO Debate." *Huffington Post.* April 21. http://www.huffingtonpost.com/2014/04/21/hawaii-gmo-flash-point_n_5187599.html.

McMichael, Philip. 2009. "A Food Regime Analysis of the 'World Food Crisis.'" *Agriculture and Human Values* 26 (4): 281–295.

Mitra, Maureen N. 2014. "Trouble in Paradise." *Earth Island Journal* 29 (1): 18–23.

Moose, Stephen P., and Rita H. Mumm. 2008. "Molecular Plant Breeding as the Foundation for 21st Century Crop Improvement." *Plant Physiology* 147 (3): 969–977.

Morris, Michael, Kate Dreher, Jean-Marcel Ribaut, and Mireille Khairallah. 2003. "Money Matters (II): Costs of Maize Inbred Line Conversion Schemes at CIMMYT Using Conventional and Marker-Assisted Selection." *Molecular Breeding* 11 (3): 235–247.

Paik, Koohan, and Jerry Mander. 2009. *The Superferry Chronicles: Hawaii's Uprising against Militarism, Commercialism, and the Desecration of the Earth.* Kihei: Koa Books.

Paiva, Derek. 1999. "Corn's Top Dogs." *Hawai'i Business.* August 1999. https://www.highbeam.com/doc/1G1-55412095.html.

Pollack, Andrew. 2013. "Unease in Hawaii's Cornfields." *New York Times.* October 7. http://www.nytimes.com/2013/10/08/business/fight-over-genetically-altered-crops-flares-in-hawaii.html.

Rice, Gareth. 2010. "Reflections on Interviewing Elites." *Area* 42 (1): 70–75.

Sakiyama, Ney Sussumu, Helaine Christine Cancela Ramos, Eveline Teixeira Caixeta, and Messias Gonzaga. 2014. "Plant Breeding with Marker-Assisted Selection in Brazil." *Crop Breeding and Applied Biotechnology* 14 (1): 54–60.

Schrire, Margot. 2011. "Monsanto Funds Education and Research at CTAHR." University of Hawai'i Foundation.

Schurman, Rachel, and William A. Munro. 2010. *Fighting for the Future of Food: Activists versus Agribusiness in the Struggle over Biotechnology.* Minneapolis: University of Minnesota Press.

Singh, Devindar. 1983. *Resource Requirements and Costs for Corn Production in Hawaii.* University of Hawai'i at Mānoa, College of Tropical Agriculture and Human Resources, Research Series.

Suarez, Adrienne I. 2004. "Avoiding the Next Hokuli'a: The Debate over Hawai'i's Agricultural Subdivisions." *University of Hawai'i Law Review* 27:441–467.

Suryanata, Krisnawati. 2000. "Products from Paradise: The Social Construction of Hawaii Crops." *Agriculture and Human Values* 17 (2): 181–189.

———. 2002. "Diversified Agriculture, Land Use, and Agrofood Networks in Hawaii." *Economic Geography* 78 (1): 71–86.

Thomassin, Paul J., PingSun Leung, and Jaw-Kai Wang. 1985. "The Economic Feasibility of Growing Taro in Rotation with Field Corn for Dairy Silage on the Island of Kauai." University of Hawai'i at Mānoa, College of Tropical Agriculture and Human Resources, Vol. 61.

Thompson, Paul B. 1995. *The Spirit of the Soil: Agriculture and Environmental Ethics.* London and New York: Routledge.

USDA, National Agricultural Statistics Service. Multiple years. "Statistics of Hawaii Agriculture." http://www.nass.usda.gov/Statistics_by_State/Hawaii/Publications/Annual_Statistical_Bulletin/.

———. 2015. "Pacific Region—Hawaii, Seed Crops." http://www.nass.usda.gov/Statistics_by_State/Hawaii/Publications/Sugarcane_and_Specialty_Crops/2015HawaiiSeedCrops.pdf.

Warman, Arturo. 2003. *Corn & Capitalism: How a Botanical Bastard Grew to Global Dominance.* Chapel Hill: University of North Carolina Press.

Xu, Yunbi, Zhi-Kang Li, and Michael J. Thomson. 2012. "Molecular Breeding in Plants: Moving into the Mainstream." *Molecular Breeding* 29 (4): 831–832.

7 | Farming on the Margin

Women Organic Farmers in Hawai'i

Aya Hirata Kimura

Women farmers are on the rise, particularly in sustainable and organic agriculture. Organic agriculture itself is a growing sector within agriculture, seeing many new entrants and rapidly expanding markets. Organic agriculture has also provided a relatively friendly space for women farmers. The proportion of women farmers in organic agriculture is higher than in conventional agriculture, and there is increasing interest among women in sustainable farming. Women have also played an important role in social movements for sustainable agriculture and food systems, spearheading initiatives for food security, reduced agrochemical use, and localization of food systems (Trauger 2004).

The relative prominence of women in the organic/sustainable agricultural movement is in contrast with the historical position of women in conventional agriculture, which has typically been dominated by men. Michael Bell and his colleagues (Bell 2004; Campbell, Bell, and Finney 2006) have suggested that organic agriculture embraces, rather than denigrates, characteristics that are conventionally marked "feminine." They theorize this as a contrast between dialogic agriculture versus monologic agriculture in which the former allows women and men to question the strict enforcement of gender stereotypes. Within the organic/sustainable agricultural community, women take leadership roles and feel welcomed and respected by fellow farmers, whereas in conventional agriculture, women often do not get respect and validation as full-fledged farmers.

This literature might seem to imply that there are few gendered challenges for women organic farmers. This chapter argues the contrary and clarifies that the external political economy surrounding organic agriculture needs to be considered in order to understand gendered dynamics in organic agriculture. Despite its less male-dominated culture in comparison with conventional agriculture, we still need to ask: what are the challenges facing organic farmers in general and how does gender complicate social relationships around organic agriculture? All organic farmers operate within a larger political economy, and none of them are entirely insulated from the economic, political, social, and cultural challenges that face organic agriculture.

Drawing on interviews with women organic farmers in Hawai'i, this chapter finds that the challenges of organic farmers are complex and multifaceted. While economic threats are important to smaller organic farmers, cultural and social challenges are no less significant. In particular, this chapter finds that such challenges stem from mainstream institutions' lingering embrace of modernization, intensification, and expansion, as well as the increasing prominence of genetically modified organisms (GMOs) in conventional agriculture. These are the areas in which organic farmers—particularly those who are philosophically committed to the original principles of organic farming—tend to differ from the conventional agricultural community.

This tension is gendered. Given the long-standing image of farmers as male, women organic farmers often find themselves in the awkward position of being considered "pseudo" farmers, or merely "gardeners." Furthermore, some women farmers who adhere to the ideals of "organic integrity" tend to be vocal in the contentious politics of GMOs. Given the social stereotype that casts women as irrational and anti-science, they need to find a delicate balance as they position themselves in such political debates. Embodying the political and cultural marginalization of organic agriculture, women organic farmers' experiences reflect the broader challenges facing organic agriculture in its defiance of the mainstream tenets that value intensification, profit maximization, and GMOs.

Women in Organic Agriculture

Organic agriculture is defined by the US Department of Agriculture (USDA) as farming that emphasizes "the use of renewable resources and the conservation of soil and water to enhance environmental quality for future generations . . . without using most conventional pesticides; fertilizers made with synthetic ingredients or sewage sludge; bioengineering; or ionizing radiation" (http://www.nal.usda.gov/afsic/pubs/ofp/ofp.shtml). First popularized in the United States by J. I. Rodale in the 1940s, the organic agriculture movement gained momentum in the 1960s, and was further institutionalized in the 2000s when the USDA set up the National Organic Program, which eventually created a framework for National Organic Certification (Guthman 2004a).

Organic farming is an important space in agriculture for multiple reasons. First, it addresses many of the environmental and health impacts of conventional agriculture. For instance, organic farms tend to have greater biodiversity (Hole et al. 2005), better water quality due to reduced pesticide and nutrient pollution, and better carbon sequestration (Greene et al. 2009). Organic produce also tends to have less pesticide residues (Baker et al. 2002; O'Riordan and Cobb 2001).

Organic agriculture is also important economically, as it has shown tremendous growth in recent years. Sales of organic food exhibited an annual growth rate of 17 percent in the last decade, and there are now 4.6 million acres under organic agriculture in the United States. In Hawai'i, there are more than 150 organic farms (USDA 2015), an increase from 100 in the 1990s (Radovich and Valenzuela 1999).

Women and organic agriculture have an interesting relationship, as there are more women in organic agriculture than in conventional agriculture. While women have made headway in various occupational sectors—now constituting 30 percent of lawyers, 32 percent of doctors, and 42 percent of accountants—according to the 2007 (and 2012) Agricultural Census, only 14 percent of principal farm operators are women, which is similar to the proportion of women in the military (Women's Memorial Foundation 2011). Agriculture has been and still is a male-dominated occupational sector.

Yet a growing number of women are finding opportunities in agriculture. A 2009 article in the *Washington Post* entitled "Female Farmers Sprouting" suggested that women farmers, though still a relatively small group, are "the changing face of American agriculture" (Aratani 2009). Although 14 percent is low, it is a significant improvement from the 1970s when only 5 percent of farmers were women. The number of farms operated by women in the United States has more than doubled since 1978, from just over 100,000 to almost 250,000 (Moskin 2005). With both primary and secondary operators combined, there are close to one million female farmers in the United States (Hoppe and Korb 2013). The situation is similar in Hawai'i. The state's farming community is facing a successor problem, with many opting for more profitable jobs, but 25 percent of the principal farmers are women (USDA 2008).

As the *Washington Post* article observed, "while men tend to run larger farms focused on such commodity crops as soybeans and wheat, women tend to run smaller, more specialized enterprises selling heirloom tomatoes and grass-fed beef to well-heeled, eco-conscious consumers" (Aratani 2009). It seems that when women decide to farm, they do it differently. Women-run farms are smaller in scale, grow more diverse crops, and tend to choose direct marketing outlets such as farmers' markets and Community Supported Agriculture (Allen 2007). A particularly salient characteristic is the preference of women farmers for organic agriculture. The Census of Agriculture indicates a higher percentage of women farmers in organic agriculture than in conventional agriculture. Nationally, women constitute 22 percent of organic farmers compared to 14 percent in general agriculture (USDA 2007).

Feminist studies of agriculture have documented how women have been subjugated and discriminated against in rural communities. The definition of a "good farmer" is resolutely masculinized (Little 2002). Historically, women have played significant roles on the farm, but oftentimes their work is not properly recognized and appreciated (Sachs 1983). Women have tended livestock and produced vegetables, but their work has frequently been considered merely a side job despite their important contributions to farm economy. They have been marginalized from decision-making roles on the farm (Sachs 1983) and important sources of agricultural knowledge such as extension services (Whatmore 1991; Brandth 1995; Trauger et al. 2008).

Scholars have found that, in contrast to conventional agriculture, organic and sustainable agriculture is more welcoming and appreciative of women. Sustainable agriculture movements in the United States have included many women activists (DeLind and Ferguson 1999). Women farmers in sustainable agricultural communities are respected as farmers, and men exhibit attitudes less propelled by dominant masculinity (Campbell, Bell, and Finney 2006). Women farmers are more likely to identify themselves as "farmers" rather than subjugating their positions as "partners" and "helpers" in sustainable agriculture than in conventional agriculture (Trauger 2004).

Hence one might conclude that gender is no longer a relevant variable in understanding the social and cultural dynamics of organic agriculture in the United States. However, feminist analyses of other industries offer a warning. They have found that when an industry becomes feminized, its wages, social protection, and prestige tend to decrease due to the devaluation of female work (Levanon, England, and Allison 2009). This suggests that a feminist analysis is still necessary to explore the politics of organic agriculture.

Hawai'i Organic Agriculture

According to USDA data, there are 166 organic farms that are certified or have exempt status (annual sales below $5,000) in Hawai'i (USDA 2015). Some older farms such as Dunlap Farm and Maui Ono Organic Farm started in the 1980s. However, many are new, and about 40 percent of the organic farmers in Hawai'i have been in the business for less than twenty years (USDA 2015).

Unlike in other states, notably California where many large-scale organic farms exist, Hawai'i organic farms tend to be small in size. Besides large ranches that raise cattle such as Beef and Bloom (345 acres, Māui) and Double D. Ranch (500 acres, Big Island), there are only a few crop farms that are bigger than twenty acres. Relatively large organic farms include Kaua'i Organic Farms (45 acres, Kaua'i), Moolaa Organica (27 acres, Kaua'i), Molokai Island Farm (25 acres, Molokai), Waianu Farm (20 acres, Big Island), and Touching the Earth (20 acres, Big Island), but they are nowhere near the size of the mega-farms in California that have several thousand acres.[1] On the island of O'ahu, the largest organic farm is MA'O Organic Farm (24 acres) in Wai'anae, which purchased additional land in 2010 with funding from the Hawai'i Community Foundation and the Omidyar Ohana Fund (Ubay 2010).

A general growth in interest and sales in the 2000s and, in particular, the entrance of Whole Foods Market in Hawai'i in 2008 have been powerful drivers in mainstreaming the concept of organic agriculture in the state. For instance, even relatively large conventional farms have expressed interest in organic agriculture. Nalo Farm's Dean Okimoto, a former president of the Hawai'i Farm Bureau Federation, was reported to be "seriously considering" going organic (Wu 2007). Organic techniques such as composting and cover crops, foreign to many conventional farmers twenty years ago, are becoming increasingly commonplace.

Given organic agriculture's sustainability and higher market values, one might be justified in expecting it to become mainstream quickly. However, looking at the larger picture, it is evident that organic agriculture in Hawai'i still straddles the borderline between marginal and mainstream. For instance, the Hawai'i Farm Bureau Federation has long maintained a skeptical stance towards organic agriculture. Since the early 2000s, some organic farmers have wanted to create their own committee within the Hawai'i Farm Bureau Federation, but it was only in 2011 that the federation allowed an organic agriculture committee to be established. Furthermore, research on organic agriculture seems to get relatively little attention within broader agricultural research and development. Public policy support is also minimal. In 2009, for instance, the Hawai'i State Legislature Senate Concurrent Resolution (SCR) 38 tried but has so far failed to obtain increased assistance for organic farmers from the Department of Agriculture.

This presents a puzzle: what are the obstacles for organic agriculture in gaining legitimacy and full support in the state and its agricultural community? What is the relationship between organic agriculture and its conventional counterpart? This chapter explores the challenges to Hawai'i organic agriculture through women farmers' experiences. Being in the marginal position of being doubly unconventional—women and organic—these farmers are well-situated to highlight the structure of marginality and exclusion in agricultural communities.

The Study

The data for this study are drawn from interviews with thirteen women organic farmers in the State of Hawai'i. I conducted in-depth semistructured interviews with women farmers on the islands of Hawai'i (the Big Island) and O'ahu. All interviews were conducted in 2011, and all were tape-recorded and later transcribed. In terms of ethnic identification, nine interviewees are white, two are Asian, one is Native American, and one is Hispanic. Their age ranged from thirties to seventies, but most of them were middle-aged. Their educational levels varied from high school dropout to PhD. Their farms are mostly small, and no one exceeded twenty-five acres. During the interviews, I asked the women about their personal backgrounds, why they chose to farm organically, and what they saw as major challenges for women organic farmers.

Women in Organic Agriculture: Experiences

The interviewees grow many different things. One woman has a dairy goat farm, and two have coffee farms, while the others grow mixed vegetables and fruits. Many of my interviewees became involved in farming not by marrying a farmer, which is the historical pattern for women in agriculture, but by starting their farms themselves, either alone or with partners.

When I asked why they chose to farm, all of their responses were about the love of growing things, being out on the farm, and the commitment to sustainability. They all described their farms as beautiful, and as I listened to them describe how much they enjoy being out on the farm, it was not hard to believe their words. Beauty and the joy of working the land were not the only motivation, however. Many explained that farming is a way for a woman to be her own boss. In many workplaces, gender discrimination is still prevalent and women face tremendous difficulties moving up a corporate career ladder. Owning their own farms and working there provided many of my interviewees the opportunity to avoid subordinating themselves to others and to enjoy autonomy while working. Two respondents used the phrase, "I am my own boss here," contrasting it with their experiences of sexist workplaces where they had worked before. While agriculture is often associated with physically demanding work that is coded masculine, and my respondents did not deny that it was a lot of hard work, for them, it was actually a safe, autonomous space.

Upon further probing into their biographies, many respondents connected farming to an escape from the patriarchal world. One woman described agriculture as a way to get away from her painful past, when she was beaten by her husband, struggled to raise two kids as a single mother, and was discriminated against in her workplace. For her, "agriculture is the best thing that has happened to me." Another woman bought her farm with the money she was awarded by a court for her workplace's discriminatory treatment of her as a woman of color. For women, who often face explicit and implicit sexism at work, farms can represent a rare space where they can be free from subjugating forces.

With regard to why they chose organic agriculture, farming conventionally was not even an option for many of them. Several women used the word "poison" to refer to agrochemicals, and when I posed the question of why organic, one responded by asking me a rhetorical question: "why would you want to use poison?" Despite the spectacular growth in organic farming, and contrary to other studies that have indicated the higher premium for organic produce as the major motivation for organic farmers, none of the interviewees mentioned economic gain as a reason to farm organically. Instead, they couched their answers in terms of commitment to the sustainability of the environment and concern about the health of their friends, their families, and themselves.[2]

When asked about challenges, the interviewees' responses resonated in many ways with the challenges identified in organic farming in general. As in other studies that have documented the problems faced by organic farmers, my interviewees discussed problems such as access to affordable land, pests and diseases, the difficulty of marketing, and lack of information and training. For instance, one woman farms on leased land, but her landlord in California has been reluctant about renewing her lease. Because organic agriculture requires much investment in building up good soil, and it also needs a three-year transition period until a farm can be certified organic, unavailability of long-term leases or affordable land is a major problem. Another challenge that she faced was the lack of access to skilled and stable labor, as organic farming is labor intensive compared to conventional farming, relying, for example, on pulling weeds rather than using herbicides. These responses echoed the results of another survey of Hawai'i organic farmers, which was conducted in 2007 (Radovich, Cox, and Hollyer 2009). Some farmers have participated in creative labor arrangements such as World Wide Opportunities on Organic Farms (WWOOF) (Mostafanezhad et al., this volume) and open up their farms to volunteer tourists who would work in exchange for on-farm accommodation.

Marketing seemed to require a diversity of approaches; the major marketing channels for the interviewees included farmers' markets, health food stores, and Community Supported Agriculture (CSA). Some were successful in getting contracts with hotels and restaurants and selling directly to customers online. Some struggled to sell their produce, particularly given the fact that selling at farmers' markets takes up time. The volatility in production levels also made it difficult for some farmers to market their produce to major retail stores or to commit to CSAs.

Others had problems with pests and diseases. When they encountered such problems, some called extension services, but many turned to books and the Internet for information and advice. Many seemed to feel isolated, in that there are not many places where they could gather to exchange information and help each other.

Above all, many of them struggled to keep their farms afloat. Organic agriculture alone was hardly sufficient, and all my interviewees combined different economic strategies to support themselves and their families. Although one interviewee made a gross income of $90,000, she was an exception; most of them earned less than $5,000 from farming itself. Given this situation, many combined different economic activities to keep their farms. Several interviewees had full-time jobs besides farming, or had retirement income to support themselves. Some worked parttime in other businesses. Healthcare was the key motivation for having other jobs. One interviewee could not afford an insurance plan, while others had insurance from a nonfarm job or through their spouses. For instance, one farmer's husband worked as a waiter at a hotel to provide the family with health insurance.

In addition to nonfarm income, many tried to add different revenue streams within agriculture. For instance, some received USDA grants to make improvements on the farm. One interviewee started offering farm tours and cooking classes using her produce on the farm. Another interviewee set up her farm as a nonprofit organization for environmental education. Similar to what Trauger (2004) found to be the case for Philadelphia women farmers, the women organic farmers I interviewed in Hawai'i had devised flexible economic strategies that combined farming and nonfarming activities.

Challenges

The interviews highlighted two additional major themes that concern organic farmers in Hawai'i. The first was the issue of GMOs, which parallels the findings of the 2007 survey of organic farmers in Hawai'i (Radovich, Cox, and Hollyer 2009). Many of my interviewees are concerned about the widespread planting of GMO crops in Hawai'i, which they fear could lead to cross-pollination and insect resistance.

In the early 1990s, Hawai'i's agriculture was at a major crossroads. The pineapple and sugarcane plantations that had dominated the island economies had been declining, and most plantations were closed by the mid-1990s. This left huge tracts of prime agricultural land available and opened fierce debates about what the future of agriculture in Hawai'i would be. While some pushed for diversified and sustainable agriculture, the seed crop industry eventually became dominant. Currently, several major seed companies such as Dow AgroSciences, Monsanto, Pioneer Hi-Bred International, Syngenta, and BASF operate farms in Hawai'i, using more than 6,600 acres, 40 percent of which is estimated to be planted with GMO crops (Gomes 2010). Hawai'i's agriculture is increasingly dependent upon this industry, as seed has become the number one crop in the state, surpassing sugar and pineapple in recent years (Schrager and Suryanata, this volume). Given the difficulty of accessing affordable and quality agricultural land in the state, the seed companies' rapid expansion is seen as taking up precious land and water that would otherwise be available for more sustainable farming.

Organic farmers, particularly women organic farmers, have been generally critical to the state's anti-GMO movement. Compared with the movement on the mainland, Hawai'i's anti-GMO movement is relatively new, but some active organizations emerged, such as Hawai'i SEED, GMO Free Kaua'i, GMO Free Maui, and GMO Free O'ahu. The issues of patenting local varieties and GMO taro have brought various Native Hawaiian groups on board as well, and they have become active participants in protesting GMOs. Often overlooked, however, is the fact that the state's anti-GMO movement was started and guided by women, including the women who farm organically. Five of my interviewees have been centrally involved in anti-GMO movements. Learning about the gradual but dramatic increase of GMO plants in the state, these women farmers felt compelled to "do something about it" and started to organize workshops and petitions.

There is a renewed vigor to the anti-GMO movement since 2013, after the interviews for this chapter were concluded. But the earlier wave of the movement was in 1998–2005, triggered by the release of GMO papaya in 1998. My interviewees were involved in that early wave, which sought various legislative measures to stop the proliferation of GMOs in the state. For instance, one interviewee successfully put forward a county resolution that banned GMO coffee on the Big Island.

Among my interviewees, there was some sense of exhaustion in regard to activism. I noticed that although GMO activism was still going on, many of the women farmers talked about it in the past tense. This "activism fatigue" might be attributed to the continuous failures of the legislature and regulatory agencies, who in their perception seem to constantly cave into industry pressure and do not provide solid oversight of GMOs. For instance, House Bill 1577 to legislate a five-year moratorium on GMO coffee in 2007 was not successful. Another bill in 2008, SB 958, originally called for a ten-year moratorium on GMO taro, was reduced to a five-year moratorium with a clause that prohibited future bans on GMOs in the state, and eventually failed to pass.

Two more factors related to the GMO issue also have to be noted. The first was the difficulty of navigating identity politics as anti-GMO activists. My interviewees were well aware of the critical gaze upon them that often cast them as "outside troublemaker ladies." The racial politics of GMOs in Hawai'i is complicated. While Native Hawaiian mobilization, especially with the issue of GMO taro, was dramatic and effective, some powerful proponents of GMOs in Hawai'i are also Native Hawaiians. For instance, both Dennis Gonsalves, the principal researcher who developed GMO papaya, and Adolph Helm, a former president of the Hawai'i Crop Improvement Association, an industry group representing the seed crop industry, are Native Hawaiians. In contrast, opponents are often depicted as nonlocal "outsiders." In criticizing the bill that proposed the ten-year moratorium on GMO taro, for instance, Helm wrote "the Hawaiian people, organizations and culture are being steered by outside influences to support Senate Bill 958" (Helm 2008). Interviewees had to carefully navigate this charged identity politics.

Second, GMOs inadvertently have reinforced the marginal position of organic agriculture. As mentioned earlier, techniques of organic agriculture have increasingly received the support of conventional farmers. Unlike several decades ago, many conventional farmers now accept the benefits of being less dependent on expensive agrochemical inputs; in addition, because of the organic sector's market growth and better profit margin, growing organic makes financial sense to many farmers. While these factors serve to reduce the historical division between organic and mainstream agriculture, the issue of GMOs emerges as another wedge. Because organic agriculture by definition bans the use of GMOs, the structural shift from plantation to biotechnology has resulted in the increasing politicization of organic agriculture in Hawai'i. Once the matter of GMOs arises, organic agriculture is back to being perceived as a threat to mainstream agriculture, especially as Hawai'i's dependence on GMO seed crops continues to grow.

As the seed business has become the "new sugar for Hawai'i" (Dicus 2011), it has, like the old plantation kings, gained significant economic and political clout of its own, and GMO skeptics that include organic farmers are increasingly marginalized. Governor Ben Cayetano (Democrat), who was elected in 1994, promoted biotechnology as the future of agriculture in Hawai'i (Boyd 2008). The Biotech Industry Organization (BIO) even gave Cayetano the "Governor of the Year" award in 2002. Hawai'i's general farming community initially viewed the biotech turn with wariness. For instance, in 1999, the Hawai'i Farm Bureau Federation expressed concern that the "sexy" issue of biotechnology obfuscated the needs of diverse farmers (Gomes 1999). However, as the seed industry's dominance increased, the federation's position started to align with it. In 2001, it elected Tom Hill from Syngenta Seed Company as its president, and has since become increasingly pro-GMO. It issued an official policy in support of GMOs, stating that GMOs

> have proven effective in controlling disease in papaya and banana crops. Their development is essential to the growth of Hawai'i's seed industry. The FDA, USDA and Hawai'i DOA have in place a set of scientifically based procedures and control for GMOs. We support the use of GMOs for the advancement and expansion of Hawai'i agriculture. (Hawai'i Farm Bureau Federation 2007, 27)

Some believe that organic agriculture and GMOs do not have to be mutually exclusive. Indeed, the official mantra of the government and mainstream farming organizations has been "coexistence." The Hawai'i Farm Bureau Federation and the state senate have taken this position, arguing that "the long term prosperity of Hawai'i's agricultural community depends significantly on diversity" and that all farmers must have the "opportunity to choose which farming practices will be [the] best" (cited in Greenway et al. 2007). However, my interviewees felt that the rhetoric of diversity and coexistence were political euphemisms. As evident in a report coauthored by one of my interviewees, their judgment was "coexistence = contamination" (Greenway et al. 2007), and hence they felt compelled to voice skepticism and opposition to GMOs.

Contrary to the "conventionalization hypothesis" (see below) that argues that organic agriculture is moving away from the progressive values and transformative potential of the earlier organic movement, my interviewees expressed a strong commitment to organic philosophy. In contrast to large-scale farmers who practice organic agriculture purely as a business, my interviewees showed commitment to what Goldberger (2011) calls "organic integrity." Perhaps less philosophically committed organic farmers—those who are attracted to organic agriculture primarily for its marketing potential—might be willing to coexist with GMOs as long as it does not hurt their business. However, as mentioned, all of my interviewees started organic farming for much more philosophical and/or environmental reasons. Hence, their resistance to the official coexistence policy tends to be strong.

But this creates a profound dilemma, as they are increasingly made aware that such actions have worked to reinforce the marginalization of organic agriculture. This is perhaps evident in the strong frustration expressed by one interviewee who said, "When we talk about GMO, it does not go anywhere. So I don't want to talk about it anymore! I do not want to bring it up anymore." This comment was even more striking given the fact that she had been long and deeply involved in the anti-GMO movement. The risks of contamination by GMOs are not only at the genetic level; GMOs have contaminated organic agriculture's politics in the state as well. Although many feel that the use of GMOs contradicts their values and the principles of organic agriculture, which is what brought them into activism, organic agriculture's resistance to GMOs is a major obstacle to attaining policy and institutional support in a state that is increasingly dependent on seed crops in the post-plantation economy. While some organic farmers might be able to embrace the coexistence approach, these women with their strong philosophical commitment to organic agriculture have had a difficult time doing so.

Hobby Farmer or Real Farmer?

Another recurring theme in the interviews was tension over the boundary between farming and gardening. Many interviewees seemed acutely aware of a possible critical gaze upon them that might not see them as real farmers. In addition to the fact that they farm in an unorthodox way (organic), their farms tend to be smaller, and thus they are pressed to combine different income-generating activities, as summarized above. Given this situation, many interviewees seemed uncertain as to how to characterize what they do, and whether they would qualify—in other people's eyes—as real farmers. When I called to set up an interview, for instance, one woman even asked me whether I really wanted to interview her. She explained, "I don't sell much, you know," and suggested that I interview owners of bigger farms. Others were apologetic when they discussed their small income from farming and how they worked off-farm as well.

The small size of the women's farms, their reliance on off-farm income, and the diversification of their economic activities tend to reinforce the stereotype that organic agriculture is not real farming, but just gardening. Among my interviewees, there was a real tension in their efforts to delineate between hobby farming and real farming and how to situate themselves in regard to that line.

This makes sense in light of other research that has documented the historical marginalization of organic agriculture that was described earlier. Organic agriculture resists the hegemonic doctrine of expansion, mechanization, and monocropping, and hence is often considered regressive and economically not viable by the mainstream agricultural community. In thinking about this charged distinction between farming and gardening, it is noteworthy that the key proponent of the US organic agriculture movement, J. I. Rodale, initially launched his magazine as *Organic Farming and Gardening*, but gave up trying to get recognition from the mainstream agricultural institutions and eventually dropped "farming" from the name of the magazine (Guthman 2004a). Organic agriculture has long lived in this subordinate space, where it is considered unworthy of being called "farming."

But this boundary drawing is also gendered. Such negative portrayals of organic agriculture as mere "gardening" overlap with gendered notions of gardening. Many women's farming activities have been historically called "gardening" and seen as subsidiary to the "farming" done by men. Women around the world often cultivate their own plots as a way to grow food for family and community consumption. Despite these plots' important nutritional, economic, and social functions, they are often seen as merely "gardens," inferior to men's plots for commercial purposes. Women organic farmers, then, are located in the marginal space left by these belittling stereotypes.

Such criticism of organic agriculture by women as pseudo farming obfuscates the fact that many US farmers are now part-time farmers and have to engage in other activities to keep their farms. Only 50 percent of US farmers are "full-time farmers," and income from off-farm sources accounts for 90 percent of US farm households' income. Only the large farms with sales over $500,000 receive most of their income from farming (USDA 2007). So there is nothing special about women organic farmers in terms of the importance of off-farm income. Yet perhaps because of the historical marginalization of women and organic agriculture, they particularly stand out, and the image that they are somehow not real farmers has become prevalent.

In Hawai'i, the long history of plantation agriculture intensifies the discrimination against small farms. Even the notion of "diversified agriculture"—non-plantation agriculture—struggled to gain legitimacy for a long time. When the sugarcane and pineapple plantations closed, it was said that "agriculture in Hawai'i is dead," although there were many farmers who were producing vegetables and fruits. Even smaller than conventional diversified agriculture, organic farms struggle to gain legitimacy.

The situation is complicated by the fact that there do exist people in Hawai'i who purchase the land that is zoned "agricultural" due to its relatively lower price and tax benefits, plant several trees, and call it a "farm." This kind of nominal farm has been criticized by social movements that try to promote agriculture and fight urban development. Delineating the "real farmers" from the "fake farmers" is hence particularly controversial in the state.

Organic Agriculture's Cultural Positioning

It is easy to think that organic has become thoroughly mainstream. Regular retailers such as Costco and Safeway carry organic products, and the natural food store Whole Foods Market posted $14 billion in sales in 2014. Having been acquired by major corporations, many organic food companies such as Horizon Dairy (now owned by Dean Foods) and Stonyfield Yogurt (now owned by Groupe Danone) have departed from their counterculture origins. Large-scale organic farms have also appeared, such as Earthbound Farm, which sources globally and controls large tracts of land.

This is a spectacular shift from the 1960s, when organic agriculture was part of the antiestablishment, counterculture movement (Belasco 1989). Organic agriculture embodied "a system of small-scale local suppliers whose direct marketing, minimal processing, and alternative forms of ownership explicitly challenged the established food system" (Guthman 2004a, 7). Over the years, mainstream agricultural organizations slowly gave more attention and respect to organic agriculture, and government policy support also increased. In 1990, the Congress passed the Organic Foods Production Act, in response to which the USDA implemented the National Organic Program in 2002. In 2000, the Agricultural Risk Protection Act first made organic farms eligible for federal crop insurance. The 2002 Farm Bill introduced a subsidy program for those converting to organic and a new system of payments for conservation practices. The 2008 Farm Bill continued these programs and also increased mandatory funding for organic programs by fivefold compared to the previous legislation (Greene et al. 2009).

The institutionalization of organic agriculture has been coupled with the increasing scale of organic farms in the United States. The larger organic farmers tend to pursue what some scholars call a "conventionalization" strategy, which mimics the economic calculations predominant in conventional agriculture (Buck, Getz, and Guthman 1997; Guthman 1998). Many point to the establishment of national organic standards as the critical shifting point when organic agriculture started to become a simple set of prohibited and allowed substances rather than an ecological philosophy. The conventionalization thesis discusses the practices of mimicking conventional farms in terms of "appropriation" and "substitution" (Buck, Getz, and Guthman 1997). Appropriation refers to the use of purchased inputs in order to have more control over farming processes. Substitution refers to post-harvest strategies to increase profit margins by engaging in value-added activities and in processing, distribution, and retailing. Conventional farms might also convert to organic, lured by the promise of higher profits. Furthermore, some organic farms have followed these strategies in pursuit of higher profits and have become bigger. A good example is Earthbound Farm, which was started by self-proclaimed hippies but now runs more than 10,000 acres internationally.

While the conventionalization thesis suggests that organic agriculture went mainstream, there are many organic farms that do not fit this picture. Indeed, the validity of the conventionalization thesis and whether the future of organic agriculture will be dominated by large-scale "organic lite" farms are fiercely debated (Guthman 2004b). It is necessary to recognize the diversity of organic farms today. On the one hand, there are conventionalized farms, many of which have converted to organic agriculture primarily because of its economic profitability, and those farms that pursue capitalist strategies to maximize profits. They follow only the "shallow" practice of organic agriculture while monocropping and purchasing farm inputs. On the other hand, there are smaller farms, sometimes called "lifestyle" or "committed" farms, that make compost on the farm and use techniques such as mulching, polycultural or mosaic planting, and green manure.

There are some indications that organic farms run by women tend to be the latter, as opposed to conventionalized, mega organic farms. Various scholars have documented gendered orientations towards agricultural sustainability (Chiappe and Flora 1998; Hall and Mogyorody 2007). For instance, Goldberger (2011) finds that women organic farmers are more likely to contribute to environmental stewardship and community vitality than male organic farmers. This chapter's findings that my interviewees entered organic agriculture for philosophical, environmental, and health reasons, rather than economic reasons, resonates with this existing body of research.

For these smaller "lifestyle" organic farmers, scholars have noted several distinct challenges. According to Guthman (2004b), the first challenge is price competition from larger organic farms. While they may use very distinct marketing channels (larger organic farms sell on contract to grower-shippers, while the smaller ones sell independently through farmers' markets, community-supported agriculture, and direct sales to retail and restaurants), eventually the larger ones can out-compete the smaller independent growers by lowering the market price over time. The second challenge is the political threat of lowering organic standards. Epitomized in the industry lobby to allow sewage sludge and irradiation in the National Organic Standard (Dupuis and Gillon 2009), the focus on production processes rather than the philosophy of organic agriculture has made the organic standards vulnerable to pressure to dilute them. Guthman also argues that the cost of land exerts significant economic pressure on any farmer to maximize return per acre. Ultimately, she argues that agribusiness involvement "amplifies already existing dynamics that constrain the ability for even the most committed organic growers to farm in more sustainable ways" (2004b, 307).

The chapter's findings suggest there are sociocultural, in addition to economic, challenges for these smaller, "deep" organic farmers. This also resonates with the existing literature that has documented the social and cultural schism between conventional farmers and organic farmers. Sociologist Michael Bell, for instance, discussed how conventional and organic farmers have not only different agricultural techniques, but also different worldviews and social relations (2004). Peer farmers and family members exert significant psychological pressure on organic farmers or farmers who consider converting to organic (Gardebroek 2006). Indeed, while the number of organic farms has increased, they are still a very small element in the overall American rural landscape.[3] Organic agriculture accounts for only less than 1 percent of total farmland in the United States.[4]

The still-pervasive skepticism towards organic was reflected in the words of Bob Stallman, the president of the American Farm Bureau Federation, which prides itself on representing the voice of American farmers, when he spoke at a 2010 national meeting of organic farmers: "A line must be drawn between our polite and respectful engagement with consumers, and how we must aggressively respond to extremists who want to drag agriculture back to the day of 40 acres and a mule." Referring to organic farmers as anti-modern, anti-technology "extremists," this statement is indicative of the social and cultural challenges facing organic farmers today (Zerbe 2010).

Furthermore, this chapter has found that the sociocultural marginality of organic agriculture intersects with gender dynamics. The overrepresentation of women in organic agriculture might work to strengthen the idea that organic agriculture is not real "farming" but only amounts to "gardening." Philosophically oriented women organic farmers have also explicitly questioned the mainstream embrace of GMOs, thereby making organic agriculture itself a politicized realm.

In short, while the ubiquity of organic products on the supermarket shelves might seem to suggest that organic agriculture has become thoroughly mainstream, the social positionality of organic farmers within the broader agricultural community is still riddled with tension and skepticism. This is particularly true for the more philosophically oriented organic farmers with smaller farms. When there is money to be made from organic, some farmers might be willing to consider it, but skepticism about whether organic is a respectable part of American agriculture remains strong in the mainstream agricultural community. While increasing price competition, partly spurred by the entry of large farms, is an important challenge faced by small organic farmers, this chapter has examined some additional social and cultural challenges and their gendered dimensions.

Conclusion

If organic agriculture is bifurcated into two—the big agribusiness type that instrumentally pursues organic agriculture for economic profit and the smaller, more philosophically committed artisanal type—then the women farmers in Hawai'i tend to engage in the latter type. Their less capitalist and more political orientation, however, places them in a vulnerable position where they are subject to gendered pressures. The women organic farmers that I interviewed expressed a strong commitment to environmental sustainability and health and emphasized the value of sharing with community. They faced major economic challenges in making farming economically viable, but creatively diversified their economic strategies. Yet an enduring image of women organic farmers as mere gardeners has seemed to legitimize the marginal position of organic agriculture in the wider agricultural community, and their anti-GMO stance in particular has been a major sticking point in their access to mainstream agricultural institutions such as farm organizations, as well as to university research and legislative support.

What lessons does this story from Hawaiʻi have for the broader theorization of organic agriculture? I began this chapter with a summary of research on organic agriculture and women farmers that suggests that organic agriculture is more affirmative of women's identities as farmers and more amenable to their leadership. However, if organic agriculture is situated in a larger politics of agriculture, the picture is more complicated.

If women are clustered in "artisan" organic agriculture with their small, diversified operations and more politicized orientations, gendered stereotypes of irrational antiscience and "side" hobby farming might work to prevent them from accessing institutionalized resources. Therefore, organic agriculture offers a profound paradox to women farmers. On the one hand, organic agriculture provides an empowering space by affirming their identity as farmers and surrounds them with a less masculinized network of fellow farmers. On the other hand, women might feel stronger pressure from negative portrayals of organic agriculture as pseudo farming and radicalism.

This study also suggests a conundrum for the theorization of the organic agricultural community at large. As summarized in this chapter, "philosophical" and "lifestyle" organic farmers are economically marginalized because they are increasingly relegated to peripheral markets, such as farmers' markets and CSAs, while the larger farms exert downward price pressures (Constance, Choi, and Lyke-Ho-Gland 2008). In addition to such economic challenges, the growing number of women in organic agriculture might exacerbate its social and political marginalization, if negative gendered stereotypes about women remain.

Notes

1. The acreage is taken from the Hawaiʻi Organic Farming Association website at http://www.hawaiiorganic.org/.

2. We should also note that significant capital might be necessary for conventional farmers to be economically successful as they rely more upon equipment, machines,

agrochemicals, and large amounts of land. This might also present a practical motivation for many women farmers to choose organic (the farms of women farmers are on average much smaller than those of male farm operators). But this factor was not mentioned by my interviewees.

3. Organic farming's share of total agricultural sales for 2009 was still low. The highest percentages were in New Hampshire, which was only 7.05 percent, and Hawai'i, which was 1.3 percent (Kostandini, Mykerezi, and Tanellari 2011). See http://ageconsearch.umn.edu/bitstream/113539/2/jaae433ip13.pdf.

4. It's notable that the share of organic agriculture is much higher in other countries such as Switzerland (11 percent in 2007), Italy (9 percent), Uruguay (over 6 percent), and the United Kingdom (over 4 percent) (*Farm Futures* 2010).

References

Allen, Patricia. 2007. *Together at the Table.* University Park: Pennsylvania State University Press.

Aratani, Lori. 2009. "Female Farmers Sprouting: More Md., Va. Women Lead Farms." *The Washington Post.* June 28. http://www.washingtonpost.com/wp-dyn/content/article/2009/06/27/AR2009062702386.html.

Baker, B. P., C. M. Benbrook, E. Groth III, and K. L. Benbrook. 2002. "Pesticide Residues in Conventional, Integrated Pest Management (IPM)-Grown and Organic Foods: Insights from Three US Data Sets." *Food Additives & Contaminants* 19 (5): 427–446.

Belasco, Warren. 1989. *Appetite for Change.* New York: Pantheon.

Bell, M. 2004. *Farming for Us All: Practical Agriculture and the Cultivation of Sustainability.* University Park: Pennsylvania State University Press.

Boyd, Robynne. 2008. "Genetically Modified Hawai'i." *Scientific American.* December. http://www.scientificamerican.com/article.cfm?id=genetically-modified-Hawai'i.

Brandth, B. 1995. "Rural Masculinity in Transition: Gender Images in Tractor Advertisements." *Journal of Rural Studies* 11 (2): 123–133.

Buck, D., C. Getz, and J. Guthman. 1997. "From Farm to Table: The Organic Vegetable Commodity Chain of Northern California." *Sociologia Ruralis* 37 (1): 3–20.

Campbell, H., M. Bell, and M. Finney. 2006. *Country Boys: Masculinity and Rural Life.* University Park: Pennsylvania State University Press.

Chiappe, M. B., and C. Butler Flora. 1998. "Gendered Elements of the Alternative Agriculture Paradigm." *Rural Sociology* 63 (3): 372–393.

"A Closer Look at Organic." 2010. *Farm Futures.* July 15. http://www.farmfutures.com/story.aspx/a/closer/look/at/organic/40128.

Constance, D. H., J. Y. Choi, and H. Lyke-Ho-Gland. 2008. "Conventionalization, Bifurcation, and Quality of Life: Certified and Non-Certified Organic Famers in Texas." *Southern Rural Sociology* 23 (1): 208–234.

DeLind, L. B., and A. E. Ferguson. 1999. "Is This a Women's Movement? The Relationship of Gender to Community-Supported Agriculture in Michigan." *Human Organization* 58 (2): 190–200.

Dicus, Howard. 2011. "Seed Capital: A Third of Hawai'i Farm Revenue." *Hawai'i News Now.* January 11. http://www.hawaiinewsnow.com/story/13827325/seed-capital-a-third-of-Hawai'i-farm-revenue?redirected=true.

Dupuis, E. M., and S. Gillon. 2009. "Alternative Modes of Governance: Organic as Civic Engagement." *Agriculture and Human Values* 26 (1): 43–56.

Gardebroek, C. 2006. "Comparing Risk Attitudes of Organic and Non-Organic Farmers with a Bayesian Random Coefficient Model." *European Review of Agricultural Economics* 33 (4): 485–510.

Goldberger, Jessica R. 2011. "Conventionalization, Civic Engagement, and the Sustainability of Organic Agriculture." *Journal of Rural Studies* 27 (3): 288–296.

Gomes, Andrew. 1999. "Ag Leaders Plan to Lobby Hard." *Pacific Business News.* January 11. http://www.bizjournals.com/pacific/stories/1999/01/11/story1.html?page=all.

———. 2010. "Isle Seed Crop Value Jumps 26 Percent." *Honolulu Star-Advertiser.* November 18. http://www.staradvertiser.com/business/20101118_Isle_seed_crop_value_jumps_26_percent.html.

Greene, C., C. Dimitri, B. H. Lin, W. McBride, L. Oberholtzer, and T. Smith. 2009. *Emerging Issues in the U.S. Organic Industry.* Washington, DC: USDA Economic Research Service.

Greenway, Una, Melanie Bondera, Vincent Mina, Richard Spiegel, Kimberly Clark, and Routh Bolomet. 2007. "Exploring Coexistence of Diverse Farming Practices: Alternative Report." http://www.hawaiiseed.org/downloads/publications-and-reports/HI-farmers-alternative-report-2007.pdf.

Guthman, J. 1998. "Regulating Meaning, Appropriating Nature: The Codification of California Organic Agriculture." *Antipode* 30 (2): 135–154.

———. 2004a. *Agrarian Dreams: The Paradox of Organic Farming in California.* Berkeley: University of California Press.

———. 2004b. "The Trouble with 'Organic Lite' in California: A Rejoinder to the 'Conventionalisation' Debate." *Sociologia Ruralis* 44 (3): 301–316.

Hall, Alan, and Veronika Mogyorody. 2007. "Organic Farming, Gender, and the Labor Process." *Rural Sociology* 72 (2): 289–316.

"Hawai'i: Background Story of the Disturbing End to SB 958—Relating to Taro | Diversidad Biológica." 2008. *INdigenous Portal.* May 10. http://www.indigenousportal.com/es/Diversidad-Biol%C3%B3gica/Hawai'i-Background-story-of-the-disturbing-end-to-SB-958-relating-to-Taro.html.

Hawai'i Farm Bureau Federation. 2007. "Hawai'i Farm Bureau Federation Policies." http://www.bigislandfarmbureau.org/uploads/HFBF_Policies_2006.pdf.

Helm, Adolph. 2008. "Sustaining Taro in New Era." *Honolulu Star-Advertiser.* March 2. http://the.honoluluadvertiser.com/article/2008/Mar/02/op/hawaii803020334.html.

Hole, D. G., A. J. Perkins, J. D. Wilson, I. H. Alexander, P. V. Grice, and A. D. Evans. 2005. "Does Organic Farming Benefit Biodiversity?" *Biological Conservation* 122 (1): 113–130.

Hoppe, Robert A., and Penni Korb. 2013. *Characteristics of Women Farm Operators and Their Farms.* USDA Economic Research Service.

Kostandini, G., E. Mykerezi, and E. Tanellari. 2011. "Viability of Organic Production in Rural Counties: County and State-Level Evidence from the United States." *Journal of Agricultural and Applied Economics* 43 (3): 443–451.

Levanon, Asaf, Paula England, and Paul Allison. 2009. "Occupational Feminization and Pay: Assessing Causal Dynamics Using 1950–2000 US Census Data." *Social Forces* 88 (2): 865–891.

Little, J. 2002. "Rural Geography: Rural Gender Identity and the Performance of Masculinity and Femininity in the Countryside." *Progress in Human Geography* 26 (5): 665–670.

Moskin, Julia. 2005. "Women Find Their Place in the Field." *New York Times*. June 1. http://www.nytimes.com.eres.library.manoa.Hawai'i.edu/2005/06/01/dining/01farm.html?_r=2.

O'Riordan, T., and D. Cobb. 2001. "Assessing the Consequences of Converting to Organic Agriculture." *Journal of Agricultural Economics* 52 (1): 22–35.

Radovich, Theodore, and H. Valenzuela. 1999. "Organic Farming: An Overview of the Organic Farming Industry in Hawai'i." *Vegetable Crops Update* 9 (1): 1–12.

Radovich, Theodore, Linda J. Cox, and James R. Hollyer. 2009. "Overview of Organic Food Crop Systems in Hawai'i." *University of Hawai'i Sustainable Agriculture* SA-3. http://scholarspace.manoa.Hawai'i.edu/handle/10125/13448.

Sachs, Carolyn. 1983. *Invisible Farmers: Women in Agricultural Production*. Totowa, NJ: Rawman and Allanheld.

Trauger, A. 2004. "'Because They Can Do the Work': Women Farmers in Sustainable Agriculture in Pennsylvania, USA." *Gender, Place & Culture* 11 (2): 289–307.

Trauger, Amy, Carolyn Sachs, Mary Barbercheck, Nancy Ellen Kiernan, Kathy Brasier, and Jill Findeis. 2008. "Agricultural Education: Gender Identity and Knowledge Exchange." *Journal of Rural Studies* 24 (4): 432–439. doi:10.1016/j.jrurstud.2008.03.007.

Ubay, Jason. 2010. "More than Just Farming." *Hawai'i Business*. November. http://www.hawaiibusiness.com/SmallBiz/November-2010/More-Than-just-Farming/.

USDA. 2007. "Census of Agriculture 2007."

———. 2008. "Census of Agriculture 2008."

———. 2012. "Census of Agriculture 2012."

———. 2015. "Organic Survey 2014."

Whatmore, S. 1991. *Farming Women: Gender, Work, and Family Enterprise*. London: Macmillan.

Women's Memorial Foundation. 2011. "Statistics on Women in the Military." http://www.ctahr.Hawai'i.edu/oc/freepubs/pdf/VCU_4_99.pdf.

Wu, Nina. 2007. "Growing Organic: The Arrival of Whole Foods Next Year Could Help Boost the Number of Certified Isle Farms." *Star Bulletin*, July 31. http://archives.starbulletin.com/2007/07/31/business/story03.html.

Zerbe, Leah. 2010. "Sustainable Farms | U.S. Farm Bureau Federation Declares War on Sustainable Food." *Rodale*. February 10. http://www.rodale.com/sustainable-farms.

MICHELLE GALIMBA

Michelle Galimba is the second-generation owner of Kuahiwi Ranch on the Big Island. The Kuahiwi is a 10,000-acre ranch, providing local beef to noted restaurants. Michelle has a graduate degree in comparative literature from UC Berkeley but came back to her family business. The narrative was written by herself in 2012.

The Place

I am looking out of the window onto the pastures and mountains of our ranch, across the valley, and down the Ka'u coastline towards the distant lava fields of Kilauea. Rain is lashing down and I am extremely happy about that. When you are a rancher you have a visceral relationship to rain, a relationship that deepens as the years go by and you learn the intricacies of the rain in your 'āina (land). When you have been through a few cycles of drought; when you have seen how shifts in vegetation and land use change how water moves across the landscape; when you go up into the forest to see the water source in all its mystery: all of these experiences give you a better understanding of what rain is, what it means, what it might be saying. It is not just water falling out of the sky. It is intelligence. And then there are all the interactions between rain and every other element of the landscape. Rain, grass, cows: that is the trinity for ranchers; and the interactions between these three elements in any particular place is the core of what we study and practice and serve, endlessly. Even a long life is not enough time to know how these three interact right now and how they might interact, not to mention everything else—soil, volcano, ocean, hawk, mouse, fire weed, petroglyph, on and on—that are woven into every day. Of course we do all this with the human economy in mind, how to get the two to mesh, and more and more we are looking at how to support as much life as possible, human and nonhuman, on the ranch.

Family

My father grew up farming, fishing, and hunting in Kaʻu. His father came here as a young teenager from the Philippines to work at the sugarcane plantation, working up the hierarchy from field worker to crane operator. My grandfather started out working for twenty-five cents a day, but he put all four of his children through college. My father discovered that he had a passion and a talent for raising livestock, so he went to Cal Poly in San Luis Obispo, and came back to Kaʻu to work for C. Brewer, which had multiple agricultural ventures at the time: sugarcane, of course, but also macadamia nuts, orange orchards, Naʻalehu Dairy, Hawaiian Ranch, even a feedlot for finishing cattle.

My mother grew up in rural upstate New York. Her father was a photographer, but they kept and milked goats and had a large kitchen garden. My mother loves horses and the outdoors, which led her eventually to my father who was working as a cowboy for Hawaiian Ranch when they met. When I was four, my father was given a position at Naʻalehu Dairy and so began a very successful career in the dairy industry. Naʻalehu Dairy was a small grass-fed dairy at the time. The cows only got feed when they came in to be milked. As time went on the dairies moved towards more confinement and feeding because it increased production. Then milk from the mainland came in and that has been nearly the end of the dairy industry. The dairies and the sugarcane plantations went out at about the same time. A lot of land became available when the sugarcane plantations folded up and that's how we got to start our ranch: repurposing former sugar fields.

We have come a long way; next year it will be twenty years since we started our ranch business, starting from scratch. Working together as a family is an amazing experience. Sometimes it can be very, very painful; when you fight it really hurts because you can't leave it behind at "the workplace." But I think it's really hard to do agriculture without those kinds of ties that go way beyond money. Maybe you don't have to have actual blood-family but I think that a family of some sort is the essential infrastructure.

The Community

I am lucky to live in a place that is beautiful and alive, almost savagely alive, but also in a community that has not been entirely shattered by the modern economy and consumer culture. And yet it's not like this landscape and this community has been "preserved" to be a particular way; it's not that we don't have our share of the best and worst of the present day. We who live here are lucky enough to be geographically isolated and deficient in such magnets to development as calm seas and white sand beaches, so that much of the recent history of development in Hawai'i has not been replicated upon this place, specifically the turn towards a visitor-driven economy. This left us economically depressed but socially and environmentally more intact. You could look at it either negatively or positively: we're so backward here that we've actually come out in the future.

The people who live here love the land as it is, passionately. All of these skills of sustainability and local food production that are being rediscovered—recycling materials, home gardening, raising livestock—have always been a part of everyday life. A few years ago, all of that was quaint, now it's fashionable. But deeper than the skills are the values, the ethos. It's hard to really define it, although, if you live here for a while, you feel it, it becomes part of your operating system, so to speak. For instance, you know almost everyone in the community by face if not by name, and most by name. You wave at each other when seeing each other every day, even if you are driving past each other on the highway. Just this, the daily affirmation of community by a small hand-gesture is a powerful thing: a salute to each other's personhood. So that becomes a habit, to respect each other and to engage with each other, to really see each other, even if you violently disagree for the moment about some issue or problem. And we do disagree, vociferously.

Actually most of the time we disagree with outsiders who come in with plans to develop this or that. That is the thing that we all have in common: a connection to the land that I can only really explain as spiritual in nature. It's not like you even have to talk about it. The land is big, it has a big presence, it has not been whittled away, domesticated, paved over, and commodified. It is the common ground, a spiritual presence, a source of wonder, pride, and identity. None of these things are measurable right now, in economic terms or any other terms, and yet what could be more important? What could be more essential to our well-being—to our "diet" in a larger sense—than all of the immaterial benefits of close connection with the land and the community?

Ranching

I am a rancher for a very simple, selfish reason: it makes me happy. Happy to be outside in the open spaces; happy to contribute to the food supply in a reasonably sustainable way; happy to work with animals, plants, rocks, water; happy to get my hands dirty and to work hard. Most farmers and ranchers do what they do for reasons that go way beyond money. Back in the nineties we used to say that we did it for the lifestyle, because that was the only way to talk about it as an entire thing. It is the truth, but only part of it. In the last few years we have been able to talk about our pride in providing food. As a society our values have changed enough to where people see producing food as being of value. A decade or two ago, to say that you produced food had no meaning to most people or a negative meaning, as in that you were too dumb to do anything else. That stigma still lingers a bit: the idea that doing manual labor is demeaning and tedious and that mental or administrative labor is more prestigious. You have to do both to be fully alive, in my opinion, to use your brain and body in a beautiful way.

To be honest, the food part is not what I value the most. What motivates me is the process of working with the land. I'm not sure, but I think this is what motivates most of us in agriculture. There might be some people for whom it is just a mechanical process, a purely economic equation. But for most of us, and especially the ranchers, the paycheck is getting to be with the land, day after day, year after year, and try to figure out how to do it better. How can we raise better animals; how can we make the land healthier and improve the pastures; how can we contribute to our communities; how do we maintain a sound business so that we can go on working with the land? Because most ranchers work at landscape scale as a matter of course, we also tend to think holistically almost without realizing it. Allan Savory has written about holistic management and how important it is to work from and towards a holistic goal: what is it that you really, ultimately want to accomplish? It is difficult to come to know what the holistic goals are, even for a very small family-run ranch like ours. Every person has their own perspective, and strong, usually defensive reasons for their perspective. But a goal starts from the very basic questions: What do you want your life to look like? How do you want it to function to everyone now and in the future?

There are some amazing women in the beef industry: my ranch "neighbors" Lani Petrie of Kapapala Ranch, MJ and Tracy Andrade of MJ Ranch, Tita Stack and Sara Moore of Kealia Ranch, as well as my beef processor Jill Mattos of Hawaii Beef Producers and RJ Ranch, to name just a few in the immediate vicinity. When people work hard together gender is not so important, so much as just being there as a person, with what you have to offer. Certain jobs are a better fit for the guys, such as wrestling with the really big calves where you need a lot of strength; and certain jobs are a better fit for the girls, such as caring for sick or orphan animals. Ranching is physically demanding and sometimes I've wished I had more physical strength and daring, but women have a lot to offer to agriculture and ranching. Our ability to think holistically and to think in terms of complex relationships is increasingly valuable as we all come to realize that we live in a closed system in which everything is related.

Going Local

Our ranch started the process of producing local beef about five years ago. It takes a little over two years to raise an animal for beef. Before that we had been exporting young animals to the mainland and having them finished there, the logic being that it is more cost-effective to ship the animals to the feed than the feed to the animals. It cost more to raise the animals here than to ship them away, finish and slaughter them there, and then ship the beef back. Also up until the nineties there was very little appreciation for the finer qualities of beef raised on grass; it was actually looked down upon as a lesser product not so long ago. However, the grass-fed market has grown, and the local food movement has grown over the years. This has intersected with the extreme volatility in the commodities market in recent years to even out the risks of producing for the local market versus shipping the calves away.

It was very much an uphill struggle to develop the capability of producing local beef for the local market, especially to get our product to the O'ahu market. There was a huge learning curve for us: everything from feed rations to labeling laws to cold-chain logistics to marketing. Luckily, we got help from some really wonderful people who believe in and support local agriculture with all they've got. The support of the chefs and food writers has been critical in creating the excitement around local food, which creates the demand for the farmers and ranchers to build their businesses around. Farmers' markets and that direct support and interaction with consumers is a vital part of the process, and taught me a great deal. There is a huge difference in the spirit of the transactions that happen at a supermarket and those at a good farmers' market, where the product is enveloped in the relationships that are created and there is information that passes from the vendors to the consumers. I'm not talking about brochures or other overt information, but rather the information that passes verbally and nonverbally that brings the consumers into the world of the producers just a little bit, and vice versa.

The Food System

It is fashionable right now to trash industrial agriculture and agribusiness as great villains but I think a lot of people in agriculture are put off and even offended by these attacks, not because we love the industrial agriculture model—we struggle with its shortcomings daily—but because the critique is simplistic and irresponsible. We all got into this mess together, as an entire society that has gone down the path of industrialization, and we shaped agriculture to the demands and values of an industrial society. My hope is that we move away from the industrial society towards a more ecological-informed society, but we need to take responsibility for the choices that our society made and learn both the good and the bad from those choices. It is important that we look at the history of agriculture squarely, soberly, and with humility. I understand that there is a process of becoming aware of what we previously took for granted, a process of changing one's perspective. I experienced that in college with feminism. At first I was very angry and hurt by the injustices I could now perceive against women, the subtle and not-so-subtle hurt that has been inflicted and I wanted to fight, I wanted all the women I knew to fight against their husbands, fathers, the patriarchy. But getting out into the world I met a lot of extremely strong women who negotiated gender roles effectively with grace and humor, and I realized that it was not so simple.

Likewise, I think it is a mistake to demonize certain kinds of agriculture wholesale, to call evil upon the entirety of industrial agriculture without seeing how it came about and how it fits with the rest of the industrial society. It's possible to celebrate and encourage more sustainable ways of producing food in a nondualistic manner; to recognize that simple black and white, good versus evil stories are not very helpful. Being a farmer or rancher, actually doing it, is very humbling; how you plan things to happen and how they actually turn out are going to be much different. You are always living with great uncertainty; that is normal and it is very humbling, as well. The most important thing is being able to see the world as it actually is and going from there towards whatever your vision might be, one step at a time.

8 | Labor of Meaning, Labor of Need

Organic Farm Volunteering in Hawai'i

MARY MOSTAFANEZHAD, KRISNAWATI SURYANATA, SALEH AZIZI, AND NICOLE MILNE

In 2010 a large human trafficking case in the United States was reported on Hawai'i Island, Hawai'i. More than four hundred Thai workers were allegedly brought to Hawai'i under false pretenses of high-paying agricultural jobs. Many of the workers paid the recruiting company, Global Horizon, between USD 9,000 to 21,000 to secure their employment. Yet, after their arrival the workers' documents were confiscated and they were forced to work at local farms for free or at lower wages than promised (Downes 2010). Honolulu immigration attorney Clare Hanusz explained in an interview with Keoki Kerr of KITV, a local TV station, on September 8, 2010: "They couldn't run away. They didn't have their documents. They were trapped. They were literally trapped." The gross human rights abuses eerily hark back to the indentured labor plantation days of the 1800s in Hawai'i (Okamura 1980). It also throws a spotlight onto the extant labor problems in the state, where global competition and high costs of production have led to the widespread decline of agricultural production since the mid-twentieth century.

In recent years, small organic farmers in Hawai'i have increasingly turned to volunteered labor to lower their operational costs. Through programs such as World Wide Opportunities on Organic Farms (WWOOF), they recruit volunteers—colloquially known as WWOOFers—who barter 15–30 hours of labor per week in exchange for food and shelter (Mostafanezhad et al. 2014; Azizi and Mostafanezhad, 2014). Such farm volunteering programs emerge out of a growing "experience economy" (Pine and Gilmore 1999), and the rising popularity of experiential and socially conscious tourism such as volunteer tourism—now the fastest growing niche tourism market in the world (Mostafanezhad 2014). Programs like WWOOF appeal especially to middle- and upper-class urban and suburban youth who often feel alienated by the practices of contemporary agro-food regimes. This backlash has led to new trends in popular culture that emphasize "back to the land" strategies for living authentically, which are materializing in the emergent forms of alternative tourism experiences.

Popular media is overwhelmingly optimistic of the growth of WWOOF as a win-win exchange between farmers and tourists (Madden 2010; Ganaden 2014). Volunteers WWOOFing in Hawai'i have the added benefit of enjoying the islands' unique landscapes while contributing to popular social movements around local food production. Hawai'i's popularity as a WWOOF destination dovetails with recent findings that WWOOF farms tend to be located "in high environmental/scenic quality locations and 'bohemian' cultural settings, but few in conventional farm regions, especially those with large farms and dominant 'modern' agricultural practices" (Yamamoto and Engelsted 2014, 264). Thus, Hawai'i may be seen as an ideal destination for WWOOF volunteers, where they may realize their desire for meaningful encounters and learn about organic food production in an idyllic tropical setting.

Farm volunteering programs also resonate with the call for increased consumer participation in food and farming initiatives (e.g., Lacy 2000; Allen et al. 2003; Feagan 2007). Proponents argue that initiatives such as farmers' markets, community-supported agriculture, community and school gardens, and food cooperatives constitute the seeds of "civic agriculture" (Lyson 2000), and create the space for increased participation in food democracy. However, Laura deLind (2002) points out that many of these laudable initiatives primarily focus on providing consumers with alternative market arrangements. She cautions that while responsible consumption is important, it is not in itself a civic activity. Along these lines, farm volunteering provides opportunities to be physically engaged in community work and helps nurture a sense of belonging to a place that has the potential to enable food citizenship and democracy.

In this chapter, we examine how farm volunteering programs such as WWOOF meet volunteer tourists' desires for authentic experiences in farming and organic farmers' needs to reduce their labor costs. Drawing on ethnographic and survey data collected from WWOOF farm hosts and volunteers on O'ahu and Hawai'i Island (known locally as "the Big Island"), we contend that the logic and rationale that propel the WWOOF movement do not address the underlying problems that have plagued small farms, especially in Hawai'i. Thus, we argue that while WWOOF provides a short-term coping strategy for some organic farmers, it is not sufficient to sustain a robust organic industry in a state where agriculture—both conventional and organic—continues to face structural disadvantages.

Organic Farming and the Labor Question

Today, independent farmers are frequently subjected to simple reproduction squeezes over which they do not have control. Industrial appropriation of production processes (Goodman et al. 1987) has left farmers increasingly dependent on purchased inputs and financial capital. This is further exacerbated in areas subject to the effects of real-estate pressure on land values such as California and Hawai'i, where the cost of fixed capital is high (Guthman 2004; Suryanata 2002). Due to Hawai'i's geographic location, producers also face the "pocket market" problem (Philipp 1953), in which the Hawai'i-grown generic produce does not have competitive edge outside Hawai'i, and virtually all produce is destined for consumption within the state. Growers do not have the option to sell their products outside Hawai'i when supplies are large enough to depress local prices. Meanwhile, advances in transportation and cooling technology allow wholesalers and distributors to source abundant varieties of fresh produce from elsewhere with virtually no delay.

Squeezed on both ends of production, most farms seek to increase their profitability by lowering labor costs. Small farms might depend on unpaid labor from family members, but a majority choose to recruit low cost labor that exploits persistent inequality in race, gender, and citizenship (Friedland et al. 1981; Mitchell 1996; Thomas 1992). The strategy depends on successful lobbying for public programs and other legislation that promote and sanction the importation of unskilled labor. Wells (1996) argues that labor-intensive operations such as strawberry farming succeed in lessening the cost-price squeeze because they take advantage of the lower cost of variable capital (labor) relative to the high cost of fixed capital (land and machinery). In Hawai'i, successive importation of workers from China, Japan, Portugal, Korea, and the Philippines formed the labor force during the plantation era. The disciplining of Hawai'i's plantation laborers was achieved by maintaining a hierarchical plantation structure and ethnic segregation (Takaki 1995; Okamura 2008). When Hawai'i became a state in 1959 the plantations had to abide by federal laws. Union organization and civil rights movements rendered exploitative means of labor control neither acceptable nor legal. Yet, "the labor question"—how to recruit and mobilize labor in ways that would allow both the farm and the workers to reproduce themselves—remains.

In Hawai'i, the structural challenges discussed above have resulted in a steady decline of the agricultural sector (Suryanata 2002). In order to survive economically, small farmers must employ strategies that add value to their products, such as switching to organic production. Organic farmers expect to receive a price premium from consumers who are willing to pay additional costs for the social and environmental values embedded in the products. While this promise held true in the early stage of the organic food industry, the price premium has not been universally sustained. A USDA study on organic prices for eighteen fruits and nineteen vegetables using 2005 data found that the average market values varied greatly and were usually less than 30 percent above the conventionally grown items (Lin et al. 2008). As the organic industry grows, large organic producers have begun dominating the market and eroding the premium values. Small independent organic farmers must develop and nurture their market niches through community-supported agriculture, networking with high-end chefs, or growing specialty crops to increase farm revenue (Guthman 2004).

Organic farming is generally more labor intensive, both in field operation and in record keeping and management. The labor question is therefore even more acute in organic agriculture. While some small farms can rely on extended family networks to meet their labor needs, access to non-household labor is critical to those that wish to grow and develop commercially. Wages in organic farming are comparable to those in conventional farming, with large farms more likely to provide above average wages and fringe benefits than smaller operations. Getz, Brown, and Shreck (2008) reveal the hidden tension regarding labor in organic agriculture by pointing out that growing organic does not always translate to fair labor practices. While workers in organic farms have reduced exposure to agricultural chemicals, their risk to other occupational hazards caused by stoop labor may be higher. They argue that the rise of organics ought to give an opportunity to address the labor question in agriculture, to assure that workers receive a living wage sufficient to support their families with dignity and comfort (Getz et al. 2008).

A growing number of farmers have turned to tourism-related ventures as a way to supplement the increasingly marginal incomes from farm production. Farmstays, farm tours, bed and breakfast, boutique products, farmers' markets, and roadside stands are now commonplace among small organic farmers (Phillip et al. 2010; Sharpley 2006). For example, following the economic restructuring in New Zealand in the 1980s, tourism was identified as a viable alternative for counterbalancing the diminishing agricultural production and services, population decline, and the general economic stresses experienced in many rural areas. McIntosh and Campbell (2001, 111) argue that "WWOOF is essentially a tourism venture that has been established in New Zealand 'to compensate for deficiencies in income.'" While many farm hosts cite social and cultural motivations for participation in WWOOF, economic reasoning is also central to the experience.

World Wide Opportunities on Organic Farms in Hawai'i

Originally referred to as "Working Weekends on Organic Farms," WWOOF was founded in the United Kingdom during the early 1970s as an exchange between farmers and urban dwellers who sought the unique leisure experience of rural living. Today the WWOOF network exists in more than one hundred countries (Choo and Jamal 2009; Mosedale 2009; WWOOF USA 2014). According to the program administrator at WWOOF Hawai'i, in 2014 there were approximately 300 registered farms—200 of which actively hosted WWOOFers—and between 2,500 and 3,000 people participated in the program (Jonathan Ziegler, personal communication). Considering that the US Census of Agriculture in 2012 counted only 184 certified organic farms in the state, we can presume that not every host farm was officially certified organic.

Information used for this chapter came from three separate investigations by the chapter's coauthors between 2012 and 2014. The first set of data was collected in November 2012 through an online questionnaire using Qualtrics software. The questionnaire was sent to all (approximately 300) registered farm hosts on the WWOOF Hawai'i website. We received forty-five responses from the online survey. The questionnaire included descriptive demographic questions as well as open-ended questions such as their motivation to host WWOOFers and the costs and benefits of hosting WWOOFers.

The second set of data was gathered in December 2012 and January 2013 on O'ahu and the Big Island using semistructured interviews with fourteen farm hosts and two additional farms that do not host volunteers. Respondents were chosen through snowball sampling where research collaborators recommended and provided contact information for additional participants (Bernard 2011, 2012). The interviews were conversational in style and were conducted face-to-face with the exception of one phone interview. Interviews were arranged by telephone and all face-to-face interviews were conducted on the farm in a private area where the researchers and farm hosts were the only individuals present. All interviews were digitally recorded and later transcribed for accuracy. The transcripts were coded for recurrent themes, which were extracted from the transcripts to develop an interpretive analysis of the findings. The farms surveyed in this phase of the study range from semicommercial (including farm-to-table) to hobby farms where income generation is not the primary objective of farming. Several farms specialized in specialty crops such as tea, coffee, and kava. Farm sizes ranged from 1 acre to 200 acres with the average farm around 11 acres. Some farmers owned their land, while others were on leaseholds from the state or private land owners. About three quarters of the farms began hosting volunteers within the past six years. Two-thirds of the farmers revealed that farming was not their primary source of income. On average, most farmers host three to four WWOOFers at one time, although these numbers range between one and twenty. The average WWOOFer stayed on the farm for two months, while the duration varied from two weeks to five years.

The third set of data focus on the volunteers' perspectives and was collected in 2013 and 2014. Thirty-four WWOOF volunteers on the farms that we had contacted in phase two agreed to fill the questionnaires that asked them to indicate and describe their background, motivations to participate, and future plans. A subset of nineteen volunteers also answered open-ended questions on their ideals and reflected on their experiences. All names in this chapter are pseudonyms to protect the privacy of our research collaborators.

Farm Hosts' Perspectives

In a recent *Hana Hou! The Magazine of Hawaiian Airlines* article, a WWOOF farm host on the Big Island explains: "We're grown up hippies, and this is the lifestyle we've always wanted" (Ganaden 2014, 128). This sentiment is widely shared by the WWOOF farm hosts in this research. Data from the online survey reveals that more than 80 percent of the farm hosts are Caucasians, a striking figure considering that Caucasians comprise only 26 percent of the population in Hawai'i. Our interviews further indicate that the farmers most likely to host WWOOFers are first-time farmers from the US mainland with less than five years of experience. This new breed of farmers is more likely to participate in organic farming and, in most cases, choose to farm because of the associated lifestyle.

The farm hosts interviewed for this research participated in various business models to meet their farming and lifestyle needs. Almost all sold fruits and vegetables at local farmers' markets, retails stores, and restaurants. Several farms participated in Community-Supported Agriculture (CSA). Two of the farmers also had a bed and breakfast on the farm that the WWOOFers helped to operate. Other farms raised funds from friends, family, and crowd sourcing methods for particular projects such as developing aquaculture. All farms were constantly refining their products and farming techniques to increase their revenues.

Our respondents consider the price premium attached to organic products as critical to compensate the high costs of farming in Hawai'i. However, Page et al. (2007) report on a number of "typical" organic farm ventures and show differentiation amongst organic farms on the Big Island with regard to their ability to capture the premium. While a few farms that consistently produced high-quality products do well in this market niche, many small organic farms have struggled and cannot cope with the high costs of farming. Furthermore, a number of farmers perceived that government involvement has only further weakened their position. As Guthman (2004) points out, from the late 1970s until the early 1990s the cultural politics of organic agriculture deterred many conventional growers in

California from converting to organic, therefore protecting organic farmers from competition. The establishment of the national Organic Foods Production Act of 1990 paved the way for business-oriented (as opposed to countercultural) certification and the number of certified organic operations has grown considerably.

The entry of large growers into the organic market has eroded the price premium attached to organic—a phenomenon that occurs nationwide. Additionally, many of the small organic farmers found that their own ability to receive organic labeling (and capture the price premium) hinges on their ability to comply with the rules and standards as set by the federal regulations. Indeed, only two out of sixteen farmers interviewed in this study were certified organic. The high level of documentation and recordkeeping required to acquire and maintain certification puts extra pressure on an already overburdened workforce. One farmer we interviewed on the Big Island expressed his frustration: "You get punished for being an organic farmer . . . you have to keep a daily record of everything you buy, where you buy it and when you use it. Basically, you've got to be a book keeper." Another farmer expressed his disappointment: "A lot of the regulations require a small farmer to pull off the same amount of red tape as a multimillion-dollar farming company." As the farmers in this study allude to, large-scale organic production in the continental United States has made it difficult for small organic farms in Hawai'i to protect their niche market.

Unable to capture the price premium, some farms seek to supplement their revenue through piecemeal grants, foundation support, and other forms of soft money. For these kinds of assistance, the farm often needs to be certified nonprofit and contributes to other aspects of the community such as education. Indeed, three of the farms that we visited were certified nonprofits, which allowed them access to crucial nonfarm income streams. For example, James explained: "Part of what has helped us is that we started out nonprofit in 2003 . . . we got money from USDA to build a hen house three years ago at the value of $20,000 . . . [we then received a] county research development [grant] for backyard chickens at the value of $4,600 and a couple of $1,000 grants." While the availability of small grants has been helpful to some farmers, others feel disenfranchised by grant-making institutions. For example, Peter, a farm host on the Big Island, explained to us: "Anytime we try to apply for a grant or anything, the only people who get the grants are the grant writers that go to college to learn and the big corporations and the so called nonprofits."

Farmers in this study invariably agreed that organic farming is hard, labor-intensive work. For example, Gerry from the Big Island remarked that farmers must be "on" all the time and attend to a variety of issues. Unlike farmers on the US mainland, farmers in Hawai'i do not have "off-seasons" as they have a 52-week growing season. The high labor requirements of organic farming put extra pressure on farmers to recruit a reliable workforce. As many are newcomers to Hawai'i, they have not yet established deep connections with the local community and hence lack access to the local labor force. Conscious of these challenges, Peter explained: "The farmers that are successful on this island . . . are usually [successful] because they have this huge extended family and everyone's working together for the good of the family. But as a small farmer, to go out and try to hire people, it's hard to find anybody that's going to work for less than $10 an hour."

Whereas farms on the US mainland have access to a large migrant labor force, Hawai'i's geographic isolation makes it more challenging for the industry to tap into low-wage labor pools. Consequently, farms in Hawai'i must seek alternative strategies to recruit labor. Migrant laborers in Hawai'i have recently found inroads into some farming industries, such as the Big Island's Kona coffee and milk operations, and larger vegetable operations on O'ahu, but they are not as prevalent as on the US mainland. Peter further highlighted the critical role of volunteers to his farm operation: "WWOOFers are really the central part of my existence; I wouldn't know what to do [without them] because we are not making enough money to hire full-time or even half-time employees." For farmers such as Peter, WWOOF is a strategy to cope with the relatively high cost of farm labor, which in 2013 averaged more than $14 per hour (US Department of Agriculture 2013). In this way, WWOOF can represent a significant financial advantage for small organic farmers and in many cases determine their economic viability.

Small organic farms attempt to capture some of the premium value by selling at farmers' markets, which allow them to sell at a venue with whatever products and quantity they have. Farmers receive personal feedbacks and appreciative comments directly from the consumers, which reinforces their goals and practices. Farmers' markets also help farmers to gain exposure, since chefs, retail and wholesale buyers, food writers, and others often visit farmers' markets to discover new farms and agricultural products. WWOOFers can bring energy and new ideas that help farmers unfamiliar with the scene to participate in farmers' markets. For example, Jack from the Big Island recalled that with the help of several WWOOFers, he was able to sell his produce at the farmers' market in Waimea: "They bought a pop-up tent and it was great, whatever cash we make over there we split." Another farmer remarked: "They have a great outlook on life and help bring our products to the farmers' market."

Like the widespread turn to rural and farm tourism, WWOOF farm hosts have similarly tapped into the tourism market. James from the Big Island explained to us that the cost-price squeeze prompted him to diversify his business plan. Price competition forced him to give up on selling his products to wholesalers. He also no longer sold at farmers' markets because the limited revenue did not justify the time commitment. Struggling to make profit, he turned to hosting visitors and volunteers as a way to cope: "To tell the honest truth, probably 80 percent of what we produce gets consumed on the farm. Our largest market for our farm produce is our interns and our guests who actually come here to have the farm experience on the farm. Their fee is paying for the food . . . [and] they are also learning to produce and process." In James' case, the farm's priority is not to produce, but to facilitate the generation of income from on-farm agritourism. His strategy is not unique; another farmer from the Big Island explained how it makes more economic sense to use the products of his farm operation to host visitors than to sell: "We do the real thing here, we grow pineapples and citrus and avocados and everything . . . I got enough food here to probably feed thirty people a day for a long time."

Farm Volunteers' Perspectives

Volunteers who participate in the WWOOF program come from a range of backgrounds. Ganaden (2014, 125) reported that

> Sixty-five percent of Hawai'i's WWOOFers are women, most of them young; most come from the Mainland, but many come from Canada, Europe, and Japan. Most are people who are at a point in their lives where they are ready to find something new . . . plenty of college students, or those who've just graduated from college, taking a break and experiencing something new before they think about the next step in their lives.

An administrator at WWOOF Hawai'i explained to us how most WWOOFers are from the United States, Canada, and Europe with a limited number from Japan and Australia. While most WWOOFers are between eighteen and twenty-four years of age, the range extends to well past sixty, as many recent retirees are increasingly jumping onto the WWOOF bandwagon (Jonathan Ziegler, personal communication). WWOOFers seem to be motivated by the broader trends in popular culture towards an increased awareness of food and food production. Indeed, the complementary discourse of volunteering to help organic farmers compete with corporate agribusiness is widely adopted in WWOOF culture in Hawai'i.

Our research indicates that volunteers tend to view WWOOFing as much more than an economic exchange. As one volunteer remarked, it is "about providing the experience of meeting new people, sharing beliefs, and enjoying each other's presence." Jack, a WWOOFer on O'ahu, elaborated: "WWOOFing is a magical experience. Of course, it is very temporary and there is no secure outcome of making any profit. But that is not the point of WWOOFing." Mark, a volunteer on O'ahu, similarly remarked: "It is the collective vision of striving for a better world through farming that makes the WWOOFing program so special, and why it makes it so memorable for me." Furthermore, the new community that the WWOOFer enters on the farm is often conceptualized as a surrogate "family." For example, Jody complimented the farm host on her farm for his ability to create a family-like atmosphere: "[The farm host] created a family and treated us all as equals, we all had a lot of responsibility on the farm and he trusted us with each task."

In addition to creating new social bonds on the farm, there is a widespread interest among WWOOFers in meeting "like-minded" people. Becky, for example, commented: "I got what I expected and had an amazing traveling experience. . . . Everybody was very friendly and I met similarly minded people." Tom, another young WWOOFer, explained that he was prepared to work hard in order to establish a meaningful relationship: "Sometimes the work was hard and my back or hands would be sore, and sometimes the work was easy and I couldn't have been more relaxed. Either way I always thought that both parties got their money's worth." Andrea added: "I don't mind working extra hours once in a while because I receive many things from the farm [referring to the way the staff treats her]."

Conversely, not feeling part of a community of like-minded counterparts or, as Fred, a WWOOFer on O'ahu, put it, "good vibes" from the farm hosts, was seen as a reason to leave the farm. Jenny, for example, explained how she was seeking a more "family like" experience, which she did not find: "Part of the reason I wanted to live and work on a farm was to experience living in a close knit community. . . . I think I missed out on that." She further explained that at least five other volunteers had left the farm because "we often felt that we were there only for our labor." She understood that the farmer was under economic pressure, but nevertheless perceived this focus on farm economics to have spoiled the WWOOF experience.

Ongoing Tension

WWOOFers are conscious of farmers' need to make profit but uncertain on how to bridge this goal with their expectations. All volunteers expressed that learning organic farming was the reason they joined WWOOF, but some did not appreciate that they were expected to do mundane tasks or activities such as helping with daily chores—cooking dinner, washing dishes, sweeping the floor, etc.—which in their perception "have nothing to do with farming." Robert, for example, remarked: "I would have liked to learn more about agriculture and why we did what we did. As a WWOOFer, I felt like we did a lot of rote tasks and repetitive labor." Another volunteer commented how the farm she worked at lacked the capacity to train WWOOFers, and focused more on farm economics than on the volunteers' experience: "While the farm talks about growing people, it uses them for free labor and holds them hostage from being able to do anything about it. You are expected to work full time, leaving no time for other activities or learning development."

Farmers are well aware of this tenuous relationship. While a vast majority of the farms appreciate the volunteered labor, acknowledging how "without the labor of WWOOFers, it would be very difficult to survive," many noted the loss of privacy and added anxiety of supervising inexperienced youths who are often in need of intensive training and supervision. Annie, a volunteer on O'ahu, remarked how she understood the farmer's position, and acknowledged that some WWOOF volunteers are not good at farming: "Green Farm is already running at a deficit and needs to make a profit. If I were the farmer, I would choose [workers] with more farming experience." She summed up this underlying tension by saying: "To put it simply, a farm can't expect newbie farmers who are just learning to till soil to work at the same pace as an experienced farmer."

In other cases, the farmer-volunteer relationship was strained because of poor communication and what the WWOOFers perceived as unclear expectations of them. One WWOOFer, for example, described the confusion about what they needed to do: "[One time] we were expected to pick five pounds of only the best-looking baby greens for Farm Fresh restaurant. [But] our initial round of harvesting was not good enough, so we had to go back and do it again." Another volunteer expressed his disappointment that their daily living conditions did not match what he had in mind when signing up for WWOOF. He explained: "I expected that we would eat [the] food that we grew, but the garden was over harvested, and we were mainly given processed Costco food." Additionally, Leo, a WWOOFer from O'ahu, explained how he left the first farm because of the farm host's inability to organize the workday: "I realized that I was not discovering the island, that I spent much of the time working or in dead hours."

This is not to say that WWOOFers were always uncomfortable with unclear expectations, as several WWOOFers explained that they were willing to do whatever was asked of them as it was all part of helping out the farm. More importantly, volunteers know that this work arrangement is only temporary—making the burden acceptable. For example, Chris explained: "For me, WWOOFing should not be a vacation. I'm there to experience the life of a farmer, if only for a few months." Frank, another volunteer, further explained: "Small farms don't have enough money to pay for labor. This is where farm volunteers come in. . . . If you end up stuck here and feel like you can't get out, you're an indentured servant not far from a slave." From a farmer's point of view, however, the rapid turnover adds to the stress of depending on volunteered labor. Candice, a farmer on the Big Island, pointed out that "volunteer labor works well when we still need to develop the farm. Once business is up and running, they are not dependable enough." Thus, the temporary nature of volunteers' commitment becomes an obstacle in enacting sustained change in the labor regime.

Community Building or Labor Coping Strategy?

The relationships between farmers and volunteers are not strictly defined in transactional terms, but instead reflect their shared commitment to organic ideals. Indeed, many host farmers identify themselves as "old hippies" who seek to share their own love for the ʻāina (land). However, beyond the explicit desire to build a community committed to alternative food production, farmers in this study were most clearly motivated to host WWOOFers because of the need for labor, even if this motivation is not overtly expressed. Farmers in this study invariably agreed that their farm and livelihood would not be viable without the help of WWOOFers. Hired labor was not financially feasible and there was far too much work to do on the farm without help. They are pressured to find alternative subsidization strategies, such as turning to volunteered labor, which fall outside of the norms of conventional agriculture production.

During the early stages of farm development, organic farm volunteering can be a win-win strategy. Indeed, it is noted elsewhere how WWOOF can provide benefits to both volunteers and farm hosts: "Volunteers do benefit from their experiences . . . [and] when they leave behind new structures or improved soil, they make permanent positive changes to farms that pay future dividends" (Terry 2014, 104). However, when farms seek to become profitable and expand their businesses, the expectations of the volunteered laborers and farm hosts begin to contradict each other. While agricultural labor turnover is notoriously porous, this is especially the case with volunteered labor. Thus, while WWOOF is a start-up strategy for small organic farm operations, as a farm becomes more professional and economically successful, the viability of WWOOF as a labor coping strategy becomes less clear.

The two more successful farms we interviewed shy away from using volunteered labor because of the limited compatibility of WWOOF with profitable farming. David, a farmer from Oʻahu, attributed his farm's success to the committed, paid workforce that is dedicated to the daily operations of the farm. He has maintained a stable workforce with very low turnover by paying them higher wages and good benefits. David can manage to give his workers a living wage because he sells his products to high-end restaurants and captures a higher price. He explained to us: "You have to make sure that [product and service] quality is consistently high, and you need a dependable workforce to do that." Tom, another successful farmer from the Big Island, argued: "If you want a reliable, skilled labor force, it's better to work with the local population." Professionalizing agriculture by creating mid-level positions and an experienced work crew requires employees who will remain in Hawaiʻi. Thus, as David's and Tom's comments indicate, while WWOOF can be a useful coping strategy for some small organic farmers, as farms develop into business enterprises, the farmer's goal of affordable labor begins to contradict with the WWOOFer's goal of meaningful experience.

Conclusion

Growing interest in participatory farm experiences has led to the increased popularity of farm volunteering programs in which food production, consumption experience, and social activism appear to seamlessly merge together. By positioning farmers and volunteers to work together, WWOOF provides opportunities for activist consumers (volunteers) to deepen their participation in food democracy. Yet, as this chapter has illustrated, volunteer farm labor is also inescapably embedded within the broader agro-food system that it seeks to provide an alternative to. Thus, while opportunities to engage in food democracy may seem apparent in WWOOF, the practice is also riddled with myriad limitations that echo long-standing tensions around food and agriculture in Hawai'i.

Organic farm volunteering appears to be a win-win exchange for farmers in search of labor and volunteer tourists in search of authentic and meaningful experiences. Yet closer examination indicates that the two goals do not fit with each other as easily. Indicative of this analysis, all of the farms in this research that host volunteers were small and several consumed much of what they produced on the farm. While this may be attractive for WWOOFers who seek meaningful experiences, the farms have indeed become so "authentic" that they are not even in the market. More importantly, the extent to which these farms can be economically viable—even with the volunteered labor—is not clear.

We argue that the underlying cultural logics that motivate the expansion of organic farm volunteering programs such as WWOOF overshadow the structural challenges of organic agriculture in Hawai'i. Turning to volunteered labor represents a short-term coping strategy for some organic farmers, but in itself does not address the ongoing structural labor problem that plagues the farm sector in Hawai'i and beyond. Furthermore, this strategy is not an option to farmers who cannot accommodate volunteered labor on their farms. In spite of these challenges, WWOOF in Hawai'i remains popular. Advocates of alternative agro-food systems are hopeful that farm volunteering produces engaged "food citizens" (cf., deLind

2002)—albeit temporarily—with impacts that may last long after they end their sojourns (see the "Peace Corps effect" in Mostafanezhad 2014). Volunteered labor can play a critical role in the establishment of small organic farms, especially among first-generation farmers. As some farm hosts acknowledge, their farms might not have survived without WWOOF labor. While these small, struggling farms may not add up substantively in the agricultural economy of the state, their continued presence provides a "performative effect" (cf., Gibson-Graham 2008), which is critical to fueling the discourse of alternative agriculture and key to the ways many citizens of Hawai'i envision food democracy.

References

Allen, Patricia, Margaret FitzSimmons, Michael Goodman, and Keith Warner. 2003. "Shifting Plates in the Agrifood Landscape: The Tectonics of Alternative Agrifood Initiatives in California." *Journal of Rural Studies* 19:61–75.

Ateljevic, Irena, and Stephen Doome. 2001. "Nowhere Left to Run: A Study of Value Boundaries and Segmentation within the Backpacker Market of New Zealand." *Consumer Psychology of Tourism, Hospitality, and Leisure* 2:169.

Azizi, Saleh, and Mary Mostafanezhad. 2014. "The Phenomenology of World Wide Opportunities on Organic Farms (WWOOF) in Hawai'i: Farm Host Perspectives." In *Rural Tourism: An International Perspective,* edited by K. Dasphcr, 134–150. Newcastle: Cambridge Scholars Publishing.

Bernard, Russell. 2011. *Research Methods in Anthropology.* Lanham, MD: Alta Mira.

———. 2012. *Social Research Methods: Qualitative and Quantitative Approaches.* Los Angeles, London, and New Delhi: SAGE.

Choo, H., and T. Jamal. 2009. "Tourism on Organic Farms in South Korea: A New Form of Ecotourism." *Journal of Sustainable Tourism* 17:431–454.

deLind, Laura B. 2002. "Place, Work and Civic Agriculture: Common Fields for Cultivation." *Agriculture and Human Values* 19:217–224.

Downes, Lawrence. 2010. "In an Ugly Human-Trafficking Case, Hawaii Forgets Itself." *New York Times.* September 21. http://www.nytimes.com/2010/09/21/opinion/21tue4.html?_r=0.

Feagan, Robert. 2007. "The Place of Food: Mapping Out the 'Local' in Local Food Systems." *Progress in Human Geography* 31:23–42.

Friedland, William H., Robert J. Thomas, and Amy E. Barton. 1981. *Manufacturing Green Gold: Capital, Labor, and Technology in the Lettuce Industry.* Cambridge: Cambridge University Press.

Fullagar, Simone, Erica Wilson, and Kevin Markwell. 2012. "Starting Slow: Thinking through Slow Mobilities and Experiences." *Slow Tourism: Experiences and Mobilities* 54:1.

Ganaden, Sonny. 2014. "Friend of the Farmer." *Hana Hou! The Magazine of Hawaiian Airlines* 16:123–129.

Getz, Christy, Sandy Brown, and Aimee Shreck. 2008. "Class Politics and Agricultural Exceptionalism in California's Organic Agriculture Movement." *Politics & Society* 36:478–507.

Gibson-Graham, Julie-Katherine. 2008. "Diverse Economies: Performative Practices for 'Other Worlds'." *Progress in Human Geography* 32 (5): 613–632.

Goodman, David, Bernardo Sorj, and John Wilkinson. 1987. *From Farming to Biotechnology: A Theory of Agro-Industrial Development.* Oxford and New York: Basil Blackwell.

Guthman, Julie. 2004. "Back to the Land: The Paradox of Organic Food Standards." *Environment and Planning A* 36:511–528.

Lacy, William B. 2000. "Empowering Communities through Public Work, Science and Local Food Systems: Revisiting Democracy and Globalization." *Rural Sociology* 65:3–26.

Lin, Biing-Hwan, Travis A. Smith, and Chung L. Huang. 2008. "Organic Premiums of US Fresh Produce." *Renewable Agriculture and Food Systems* 23:208–216.

Lyson, Thomas A. 2000. "Moving toward Civic Agriculture." *Choices: The Magazine of Food, Farm & Resource Issues* 15:42.

MacCannell, Dean. 1973. "Staged Authenticity: Arrangements of Social Space in Tourist Settings." *The American Journal of Sociology* 79:589–603.

Madden, Jacob. 2010. "WWOOF Your Way around the World!" CNN. http://edition.cnn.com/2010/TRAVEL/06/15/wwoofing.volunteer.farming/index.html.

McIntosh, A., and T. Campbell. 2001. "Willing Workers on Organic Farms (WWOOF): A Neglected Aspect of Farm Tourism in New Zealand." *Journal of Sustainable Tourism* 9:111–127.

Mitchell, Don. 1996. *The Lie of the Land Migrant Workers and the California Landscape.* Minneapolis: University of Minnesota Press.

Mosedale, Jan. 2009. "Wwoofing in New Zealand as Alternative Mobility and Lifestyle." *Pacific News* 32:25–27.

Mostafanezhad, Mary. 2012. "The Geography of Compassion in Volunteer Tourism." *Tourism Geographies* 15 (2): 318–337.

———. 2014. *Volunteer Tourism: Popular Humanitarianism in Neoliberal Times.* London: Ashgate.

Mostafanezhad, Mary, Saleh Azizi, and Kelsey Johansen. 2014. "Valuing Organic Farm Volunteer Tourists in Hawai'i: Farm Host Perspectives." *Current Issues in Tourism* 7:1–5.

Okamura, Jonathan. 2008. *Ethnicity and Inequality in Hawai'i.* Philadelphia, PA: Temple University Press.

Page, Christina, Lionel Bony, and Laura Schewel. 2007. "Island of Hawaii Whole System Project Phase I Report." Boulder, CO and Kamuela, HI: Rocky Mountain Institute and the Kohala Center.

Philipp, Perry F. 1953. *Diversified Agriculture of Hawaii.* Honolulu: University of Hawaii Press.

Phillip, Sharon, Colin Hunter, and Kirsty Blackstock. 2010. "A Typology for Defining Agritourism." *Tourism Management* 31:754–758.

Pine, B. Joseph, and James H. Gilmore. 1999. *The Experience Economy: Work Is Theatre and Every Business a Stage.* Cambridge, MA: Harvard Business Press.

Sharpley, Richard. 2006. "Tourism, Farming and Diversification: An Attitudinal Study." *Tourism Management* 27:1040–1052.

Suryanata, Krisnawati. 2002. "Diversified Agriculture, Land Use and Agro-Food Networks in Hawai'i." *Economic Geography* 78:71–86.

Takaki, Ronald T. 1995. *Pau Hana: Plantation Life and Labor in Hawaii, 1835–1920.* Honolulu: University of Hawai'i Press.

Terry, William. 2014. "Solving Labor Problems and Building Capacity in Sustainable Agriculture through Volunteer Tourism." *Annals of Tourism Research* 49:94–107.

Thomas, Robert J. 1992. *Citizenship, Gender and Work: Social Organization of Industrial Agriculture.* Berkeley: University of California Press.

US Department of Agriculture. 2013. "Hawaii Agricultural Labor." Honolulu: National Agricultural Statistics Service. May 28. http://www.nass.usda.gov/Statistics_by_State/Hawaii/Publications/Miscellaneous/aglaborFF.pdf. Accessed July 30, 2014.

Wells, Miriam J. 1996. *Strawberry Fields: Politics, Class and Work in California Agriculture.* Ithaca, NY and London: Cornell University Press.

WWOOF USA. 2014. https://wwoofusa.org/about/. Accessed March 16, 2014.

Yamamoto, Daisaku, and A. Katrina Engelsted. 2014. "World Wide Opportunities on Organic Farms (WWOOF) in the United States: Locations and Motivations of Volunteer Tourism Host Farms." *Journal of Sustainable Tourism* 22:964–982.

CHRIS ROBB

Robb Farms, located in Waimea on Hawai'i Island, specializes in growing US Department of Agriculture–certified organic vegetables and fruits. The owner-operator, Chris Robb, has been growing organically for over thirty years. This narrative is written by Nicole Milne, based on an interview in July 2012 with Chris Robb.

Robb Farms

Born and raised on O'ahu, Robb attended the College of Tropical Agriculture and Human Resources at the University of Hawai'i at Mānoa, and studied to become a horticulturist. Robb became interested in organic gardening when he was fifteen years old after reading the work of J. I. Rodale. Around that time he recalled seeing a steaming pile of chipped Christmas berry and thinking, "[composting] works!" Shortly thereafter he purchased a subscription to an organic gardening magazine and quickly became convinced of the need to feed soil microbes to improve the health of the soil. "The concept of organic agriculture is so simple," notes Robb, "that is what allures most people; modern culture has a tendency to dissect everything that we lose the simplicity of it all."

Following graduation Robb took a job first as the horticulturist for Watanabe Floral in Hawai'i Island's ranch town of Waimea, before shifting to work at MacFarms in the late 1980s. It was in his position at MacFarms where Robb was exposed to low-input sustainable agriculture, and began to experiment with large-scale organic production. Robb attributes his success in organic agriculture to years of reading and experimenting with organic production methods, but it was the five-year experience at MacFarms that convinced him that organic agriculture could work. Robb soon started his own farm, producing organic lettuce outside of Kailua-Kona, where he provided local markets with fresh, sustainably produced specialty lettuces. When Robb's sons went off to college and he lost a valuable source of family farm labor, he considered putting the farm on the market and going into landscaping. Growing lettuce in the hot Kailua-Kona weather was challenging, and Robb was ready for a change. However, when the ideal lot opened up in Waimea in 2004, which would provide him the ability to grow organically in weather conditions more suitable for leafy greens and other temperature sensitive vegetables, he signed the lease.

Today, over twenty years since his position at Watanabe's, Robb owns and operates Robb Farms, a fourteen-acre organic farm, just two miles from the location of his first job. Robb specializes in a variety of lettuces including baby romaine and green and red leaf, beets, fennel, leeks, Yukon gold potatoes, and broccoli and squash seasonally. Consumers can find Robb Farms produce in several local markets including Whole Foods Market and KTA Superstores.

Agricultural Labor in Hawai'i

Characterizing farm labor in Hawai'i is challenging, as most labor systems are structured in relation to the crop production system, including planting and harvesting schedules. Producers of Hawai'i's commodity crops such as coffee, macadamia nuts, and papaya typically employ fewer full-time workers and rely on part-time laborers during busy planting and harvesting periods. Some commodity producers structure wages by the task (i.e., bucket of product harvested), creating competition for wages, while providing little in terms of job security or employee benefits. Industries that continue to rely heavily on hand harvesting methods, such as coffee and macadamia nuts, are more dependent on foreign laborers and individuals from Hawai'i's immigrant communities. Robb originally operated a farm in Kailua-Kona, on Hawai'i Island's western coast, where he became familiar with the island's coffee industry.

> When Latinos came to Kona that was the best thing that ever happened to [Hawai'i's coffee] industry. . . . You get a big crew of Latinos in there and they

> are here to make money. They are making $150–$200 a day in cash. . . . They go back to Mexico and live [well]. . . . but they weren't interested in lettuce.

For Hawai'i residents, part-time agricultural labor contributes to their household income but cannot provide living wages that would allow workers to adequately support a family. Small to midsized vegetable farms often provide greater opportunity for full-time employment due to the nature of the production system. Many farmers in Hawai'i rely on their own family members while developing their agricultural businesses, and then gradually hire workers as production and sales expand. Initially Robb struggled to find workers to help on the farm.

> The first three years were hard. I'd run an ad and you might get four or five people responding and if two of them showed up to the farm I'd hire them on the spot. And I'd be lucky if they lasted a week or two.

Finding people who are serious about working on the farm has been important to Robb, as opposed to people who are just "looking for a job." When a farm operator hires someone from another professional background who has been unemployed for some time, the operator is at risk of losing that employee as soon as a position opens up in their desired industry. He says, "as soon as the next construction job comes along they're gone." The farm employs several people of Micronesian ancestry: "Because of their English skills they are [disadvantaged in Hawai'i's economy]. This [job] is security for them. I have one employee that would work seven days a week if I let him."

As Robb Farms has grown, finding labor has become easier and retaining a steady workforce more feasible. Emphasizing the importance of workers, Robb says, "if you want to be consistent in the market place, which is crucial, you have to have a reliable, trained labor force . . . and if you want a reliable, skilled labor force, it's better to work with the local population." Professionalizing agriculture by creating mid-level positions and an experienced work crew requires employees that will remain in Hawai'i. Commenting on volunteer agricultural labor programs that have become a popular solution to high labor costs and low labor availability in Hawai'i, Robb remarks:

> [Volunteers] tend to be here for the interim. What we want to create is a stable society, and I don't see [volunteers] as conducive to a successful business

> strategy. . . . With [volunteer laborers] you have to give them a place to stay and feed them, and it becomes more of a lifestyle thing than a business. At the end of the day I want to say goodbye to my workers.

It has taken time, but Robb has developed a dedicated and skilled team of workers on his farm, and he appreciates the opportunity to listen to his workers' ideas on the ways to improve the farm's operations. Farmers, notes Robb, have a tendency to fall into patterns of operation and resist change, but there is always room for improvement. Right now Robb has "great crew who work well together." Robb's sons are returning to work on the farm and his wife is moving into the position of warehouse manager.

Robb Farms provides year-round employment to approximately seven full-time workers, in addition to 100 percent full medical coverage. He contracts a locally owned human resources company to manage his workers' benefits, an expense that is the farm's biggest challenge. While farm labor constitutes a large percentage of Robb's operating expenses, he is comforted in knowing that his workers are legitimately covered and provided for. When asked about his secret to retaining workers, Robb replied:

> The health benefits are obviously a bonus, because most small business people want to go part time and stay under the threshold. Work is reliable; I always have work for them. They always get paid and the checks don't bounce (laughter). They get extra food, so hopefully their diets have gotten better as a result. . . . but you have to be up-and-up with your workers, and listen to what they have to say.

Robb continues to see access to agricultural land as a challenge to developing the state's agricultural economy. In his view, removing the speculative nature from agricultural land and ensuring that agriculturally zoned lands are used for agricultural purposes are among the first steps towards building a strong agricultural base in our rural communities. It is also important to keep food dollars circulating in our local economies. Hawai'i residents are vulnerable, says Robb, because of the amount of money we spend on imported food. "If the ship stops coming we only have roughly a week's worth of food on the island; then we will realize the importance of investing in local food production by supporting Hawai'i's farmers and purchasing locally grown food."

DEAN OKIMOTO

Dean Okimoto is a farmer and president of Nalo Farms. Established as a family business in 1983, Nalo Farms initially specialized in local fruits, daikon, and green onions. A fortunate meeting with Chef Roy Yamaguchi resulted in the creation of "Nalo Greens," a signature mix of baby greens and edible nasturtium flowers. Today, over one hundred restaurants and farmers' markets carry Nalo Farms products. This narrative is written by Nicole Milne, based on an interview with Dean Okimoto in July 2012.

Retaining a Reliable Workforce

Building an agricultural business often requires a considerable level of self-exploitation by the owner-operator in order to initiate and sustain production, and build strong relationships with buyers. Farm owners oftentimes work around the clock to ensure that planting and harvesting schedules are met and customers are provided with high-quality products upon demand. As one of the first farmers on O'ahu to market directly to local restaurants, Dean began to carve out a reputation among O'ahu's chefs early on in his career, working hard to produce a consistent, high-quality product and build what he calls a "business philosophy of service."

Hawai'i's agricultural labor needs are intimately linked to the seasonal character of the products. For example, producers of coffee or macadamia nuts rely heavily on temporary labor pools, and experience difficulties finding sufficient workers when the demand for labor peaks during harvest periods. Dean recalls the experiences of some coffee and papaya farmers who were unable to harvest their crop because of the lack of seasonal labor. Most workers prefer full-time employment so they can adequately provide for themselves and their families.

By contrast, Hawai'i's vegetable farmers operate year-round and therefore must cultivate a committed workforce that is dedicated to the daily operations and success of the farm. Dean takes pride in the 'ohana (family) they have created at Nalo Farms, their team of employees, who contribute to the vitality of the business. His employees consist of a range of skilled individuals who work in a variety of positions, including management, distribution, processing, and in the fields. According to Dean, the key to his success has been understanding the needs of his employees:

> When I first started farming, I used to lose sleep when big storms were coming through. . . . Now I can sleep through the storms; it's after the storm that I don't sleep, because you have to try and figure out how to keep your workers working, because you have to have them, but you have no income coming in . . . those are the hard parts.

He recalls that during these challenging times,

> [upper-level management] have gone as far as not taking paychecks to make sure that [the workers] get paid; my management staff has all done that because at the end of the day they see a bigger picture, that we'll all get taken care of, that we'll all be OK, as long as we stick together as a family.

He acknowledges that providing workers with this "family treatment" is possible for small to midsized producers such as Nalo Farms, which employs approximately twenty-four full-time workers. "I can give saimin [a noodle soup] to workers who have spent all day harvesting in the rain, to ensure that they have something warm to eat. Providing this to 200 workers on some of O'ahu's larger farms would be more challenging." However, Dean believes that providing these informal benefits—such as meals and social gatherings—for laborers on larger farms is still possible, and it is critical to showing appreciation for the workers. "If any farmer thinks that they did it all," says Dean, "they're wrong, because it isn't about them . . . without [the workers], there is no Nalo Farms." An upper-level farm manager, participating in the interview, commented on the pride that Nalo Farms workers have for their work:

> They are proud of what they do. What we do here, what Dean has created, is a family. . . . In the community we have senior citizens, children groups, and schools coming here . . . and they're coming here to see [Nalo Farms employees]. So [Nalo Farms employees] are not only earning a living, but the way in which you're doing it, you're proud of it, you wouldn't want to do anything different. And that's how you retain [employees].

Paying laborers higher wages and providing them with incentives—such as social gatherings, healthy food, full medical coverage, and performance-based raises—are uncommon practices in agriculture. But Dean believes they have contributed to the creation of Nalo Farms' dedicated workforce. "We get a lot of press, a lot of kudos," says Dean, "which has instilled in [our staff] a sense of pride, so it's cool for them to say they work at Nalo Farms." Worker turnover at Nalo Farms is very low, which Dean attributes to the sense of family they have created on the farm, and the benefits they provide to their workers.

When farmers have difficulties affording labor, notes Dean, it is because they price their products too low. Dean believes that products from Hawai'i farms have historically had a hard time competing against imported products, because the islands' limited refrigerated storage delivery systems reduce the products' shelf life. "But if you make sure that [product] quality is always high, then you can increase the price, expand the profit margin, and reinvest in the labor force."

The majority of Hawai'i's farms secure labor force by employing family members and new migrants at lower wages, or turning to volunteer labor (Mostafanezhad et al., this volume). In Dean's view, such subsidizations hide the real costs associated with farm labor. Consequently, consumers remain ignorant to the true cost of food production, and farmers who provide their workforce with living wages and benefits must compete with those who do not. "At some point," says Dean, "[Hawai'i's farmers] need to raise the level of what we pay agricultural laborers [in order] to keep them, and [consumers] need to get more realistic about the price of their food."

Cultivating Hawai'i's Next Generation of Farmers

Farmers must invest time and energy into educating their workforce. Workers can expand their knowledge of farm's operations if they have wide-ranging job responsibilities. An existing farmer can serve as a mentor to those interested in learning about agriculture, or in becoming a farmer. Dean himself has mentored a handful of employees, working with them to develop their knowledge of agricultural production, processing, and marketing; and instilling in them the understanding that agriculture is ultimately about culture and community. For example, a former Nalo Farms employee left in 2011 to start his own vegetable farm in Pa'auilo on Hawai'i Island. In reference to this past employee, Dean says, "I've only heard glowing reports about his produce. . . . I've heard he grows the best tomatoes." Furthermore, Dean encourages his farmworkers to value business qualities beyond their ability to grow food.

> I tell people that you have got to think of this, when you're in agriculture, as a business that is like any other business. So you have to have the ability to communicate; the ability to get along with people, on a broader range. You need to be able to understand what they need, especially when you're dealing with restaurants.

While the fastest way to increase the state's food production, according to Dean, is to encourage large established farmers to acquire additional land and ramp up production, there remains a role for the small farmer. "Small farmers are going to take a longer time to [contribute to an increase in food production] and it will be harder for them to do it," says Dean, "but [agricultural] co-ops are going to be really key in going forward [for the smaller farmers]." Additionally, Dean stresses to his employees the importance of communication—between producers, wholesalers, and buyers, including consumers—and believes it to be one of the most significant pieces missing from Hawai'i's food system. Building better business relationships and expanding our understanding of the whole agricultural system will help ensure the next generation of farmers increased success.

Nalo Farms' processing facility and farmers' markets are run largely by one of the farm's youngest employees, at age twenty-two. "Everybody wants to knock the young kids in today's world," says Dean, "and says they don't want to work hard, but this guy is unbelievable." Creating these opportunities for the upcoming generation of farmers will help develop the state's diversified agricultural economy. To create an entrepreneurial farm, says Dean, "an ability to think outside the box and look at things in different ways is key," and working with the younger generation opens the door to new business strategies.

9 | Epilogue

Aya Hirata Kimura and
Krisnawati Suryanata

Like other initiatives around the world that aspire to rebuild domestic agriculture and lessen the dependency on global food regime (McMichael 2014), citizens of Hawai'i have explored a broad range of efforts to rebuild a food system that is rooted in place and engages the local community—from small-scale initiatives such as community-supported agriculture, farmers' markets, or school gardens, to broader citizen movements and policy initiatives. As the case studies in this book attest, however, they are confronted by numerous contradictions that arise from their position in the broader political economy. We argue that in spite of the challenges, these initiatives exemplify what Gibson-Graham (2008) and others refer to as diverse economies, which are a "huge variety of economic transactions, labor practices, and economic organizations that contribute to social well-being worldwide" (Gibson-Graham 2008, 615). While they seldom are seen as a source of dynamism, they engage more people and account for more hours worked than the capitalist economy, and remain prevalent worldwide as they do in Hawai'i. In our endeavor to envision the paths towards food democracy, the volume has examined these initiatives to broaden our conception of food issues beyond a matter of food production, but as a matter of politics that is fundamentally about social, cultural, and economic powers.

While this book is about a specific place, we believe that the relevance of the book extends beyond Hawai'i, and suggest four important issues that emerged from the volume as a whole that bear importance to broader theorization in agro-food studies. First is the issue of food localization and how its class and race biases need to be considered. As an island state, Hawai'i's emphasis on local self-sufficiency is understandable and has an intuitive appeal. Hawai'i's island geography lends itself easily to the localization argument and has galvanized much support for an alternative agrofood system. Even in such a place, the book provides a cautionary note to the localization projects that can backfire in its operationalization. As Kent (this volume) argues, an increase in self-sufficiency rate is not necessarily the same as food security for all people, and policies promoting local food need a nuanced look particularly in regard to race and class biases. A narrow focus on food localization also risks advantaging the productionist approach that privileges the industrial agriculture and a small portion of farms that are adept at niche marketing over small diversified farms (Mironesco, this volume). Qualitative difference in food/farming method such as reduced environmental pollutants, protection of biodiversity, and ethics of care (Kimura, this volume) would likely be invisible if agriculture is to be judged solely on production criteria.

Second, as a state that has strong native communities, Hawai'i's experience suggests the need to integrate the sociocultural significance of food into the debate on food systems, to include discussions on how communities creatively use food as a medium for community building, cultural rejuvenation, and educational opportunities. Here and elsewhere, debates on food tend to be pulled towards the economics, sidestepping sociocultural considerations. Questions tend to boil down to "does this make economic sense?" For instance, the merit of food localization is argued on the basis that it is good for the state economy (Kent, this volume); the litmus test for organic agriculture is whether they are "real farms" with sizable output and sales, not their contribution to community education and women's empowerment (Kimura, this volume).

The dominance of neoliberal and scientific viewpoints in food-related controversies has been observed in many other instances (Moore et al. 2011; Kimura 2013). Scholars have pointed out how reducing food controversies to economic contribution and/or scientifically measurable indicators proves highly problematic, as some issues are not commensurable to monetary valuation and are riddled with uncertainty even within expert communities. At the same time, communities are making a case for appreciating cultural meanings and values of food. Taro field and fishpond restorations (Kame'eleihiwa, this volume; Kawelo, this volume) might not amount to much in terms of economic contribution or the increase in self-sufficiency rate, but they promote cultural rejuvenation, community building, a sense of pride among youth, and provide environmental education opportunities. Reframing our inquiry along this line means asking a hitherto marginalized question—"how does this policy/project nurture healthy and active communities?"

Third, Hawai'i offers excellent case studies to examine the development of genetically modified (GM) crops, and the ensuing contestation. The development of GM papayas did not follow the path of privately developed-and-owned GM seeds that typically characterizes the genetically modified organism (GMO) industry. In spite of the opposition voiced by some organic papaya growers, the anti-GM movement that was mobilized around GM papayas was relatively muted. By contrast, the seed corn industry's growth has followed a more familiar script and was fueled by large capital investment of global seed corporations. It has also generated much stronger opposition, with several legislative actions successfully taken at the county level. For instance, in 2013–2014 three counties (Kaua'i, Maui, and Hawai'i) passed bills that aim to curb the operations of large seed corn companies. While the resulting county ordinances were later invalidated in federal courts on procedural grounds, the legislative success galvanizes the anti-GMO movement. The chapters by Akatsuka (this volume) and by Schrager and Suryanata (this volume) highlight the importance of situating the local adoption and opposition of GM crops in the broader political economy, and urge evaluating the industry beyond the binary GMO/non-GMO debate. In Hawai'i, the GMO debate has been narrowly framed around seed companies' agronomic practice in their nurseries. Other important concerns such as economic alienation and corporate dominance over the landscape are subordinated to the technical discussion of the levels of pesticide use and the presence of GM crops (Schrager and Suryanata, this volume).

Fourth, experiences in Hawai'i give valuable insights to the place of local food in tourism-dependent places. No longer limited to restaurants and hotels, tourists are also interested in more intimate encounters with foods at places like farmers' markets and seek unique experiences through farm volunteering programs such as World Wide Opportunities on Organic Farms (WWOOF). Hawai'i has long struggled in negotiating its dependence on tourism economy on the one hand, and the perceived commodification and denigration of local cultures on the other. Such tension is also being played out in the encounters at local farmers' markets and organic farms.

Mostafanezhad et al. (this volume) reveal the contradictions of relying on market-based activism to build a robust agricultural economy. While volunteered labor is a critical resource for small start-up organic farms, it is not necessarily a dependable workforce. More importantly, these programs may instead contribute to the perpetuation of extant problems that plague small organic farmers in Hawai'i and beyond. Furthermore, Mironesco (this volume) argues that the strategy of relying on tourism market is a double-edged sword. While tourism increases profit potentials for local farmers and food-related industries, the production and marketing of "local" Hawai'i foods for tourists raises the questions of product accessibility to low-income local residents (see also Costa and Besio 2011).

The book presents a collection of analyses of issues around agriculture and food in Hawai'i. While each chapter is informed by a different theoretical orientation, they share a common desire to illuminate the various ways citizens of Hawai'i could participate in food democracy. Food democracy is built more on the principles of interdependence than competition; diversity than dominance; social well-being and environmental protection than economic profits as goals; collaboration and collective activism than atomized individualism. At the same time, we recognize the need to confront the understandings of capitalism that stood in the way. As such, the chapters in this volume critically analyze the different initiatives within the political economy context, identifying the pitfalls and contradictions that might threaten the effectiveness of the initiatives, while highlighting the diverse potentials to engage citizens in food democracy.

References

Costa, LeeRay, and Kathryn Besio. 2011. "Eating Hawai'i: Local Foods and Place-Making in Hawai'i Regional Cuisine." *Social & Cultural Geography* 12 (8): 839–854.

Gibson-Graham, Julie-Kathryn. 2008. "Diverse Economies: Performative Practices for 'Other Worlds.'" *Progress in Human Geography* 32 (5): 613–632.

Kimura, Aya H. 2013. *Hidden Hunger: Gender and the Politics of Smarter Foods*. Ithaca, NY: Cornell University Press.

McMichael, Philip. 2014. "Historicizing Food Sovereignty." *The Journal of Peasant Studies* 41 (6): 933–957.

Moore, Kelly, Daniel Lee Kleinman, David Hess, and Scott Frickel. 2011. "Science and Neoliberal Globalization: A Political Sociological Approach." *Theory and Society* 40 (5): 505–553.

Contributors

Neal K. Adolph Akatsuka is coordinator of publications and programs at the Mahindra Humanities Center at Harvard. He holds a master's degree in social anthropology from Harvard University. His research focuses on consumer and activist perceptions of genetically modified (GM) crops and flowers in Japan and the United States.

Saleh Azizi is a PhD candidate at the Department of Urban and Regional Planning at the University of Hawai'i at Mānoa, and an artisan cheesemaker. He currently works at the Kahumana Community in Hawai'i.

Lilikalā K. Kame'eleihiwa is professor and director at the Kamakakūokalani Center for Hawaiian Studies at the University of Hawai'i at Mānoa. She is the author of *Native Land and Foreign Desires: Pehea Lā E Pono Ai?* (1992) and numerous publications on Hawaiian ancestral knowledge, history, and cultural traditions, especially in the management of land and water resources.

George Kent is professor emeritus of political science at the University of Hawai'i. His work focuses on human rights, international relations, peace, and development, with a special focus on nutrition and children. He is the author of several books including *Freedom from Want: The Human Right to Adequate Food* (2005), *Ending Hunger Worldwide* (2011), and numerous other publications on food policies.

Aya Hirata Kimura is associate professor of women's studies at the University of Hawai'i at Mānoa. She is the author of an award-winning book,

Hidden Hunger: Gender and Politics of Smarter Food (2013), and has published on the issues of food education, farming and disaster, food standards, and consumer cooperatives.

Kem Lowry is professor emeritus and former chair of the Department of Urban and Regional Planning, former director of the Program on Conflict Resolution, University of Hawaiʻi at Mānoa, and adjunct senior fellow at the East-West Center. He has written numerous articles on issues related to land use policy, community-based resource management, and conflict resolution and evaluation.

Nicole Milne is a PhD candidate in geography at the University of Hawaiʻi at Mānoa. Her research focuses on the viability of farming in Hawaiʻi's post-plantation landscape. She is the associate vice president for programs at The Kohala Center, a nonprofit community-based center for research, education, and economic development on Hawaiʻi Island. Through her work with Hawaiʻi's farmers and ranchers, she assists clients in business development strategies, including business planning and sourcing capital to start new agricultural enterprises and expand on existing operations.

Monique Mironesco is associate professor of political science at the University of Hawaiʻi—West Oʻahu (UHWO). Her research focuses on farmers' markets in Europe and Hawaiʻi. She is a founding faculty member of the Sustainable Community Food Systems concentration at UHWO. Among a myriad of other projects, she has been teaching a service learning–centered course on the politics of food at UHWO since 2007, creating community partnerships and fostering her students' civic engagement surrounding food system issues in Hawaiʻi.

Mary Mostafanezhad is assistant professor of geography at the University of Hawaiʻi at Mānoa. Her research interests lie at the intersection of critical geopolitics and mobilities such as tourism, development, and humanitarianism. She is the author of *Volunteer Tourism: Popular Humanitarianism in Neoliberal Times* (2013), and coeditor of *Moral Encounters in Tourism* (2014) and *Cultural Encounters: Ethnographic Updates from Asia and the Pacific Islands* (2015).

Benjamin Schrager is a PhD candidate in geography at the University of Hawai'i at Mānoa. His work investigates evolving relationships in the production and consumption of food. He has researched the political economy of industrial corn in Iowa and Hawai'i. His PhD dissertation examines the cultural politics of chicken broiler industry in southern Kyushu, Japan.

Krisnawati Suryanata is associate professor of geography at the University of Hawai'i at Mānoa. Her work utilizes political ecology perspective to examine the globalization of agro-food systems, rural development, and community-based resource management. In Hawai'i, she has researched and published articles on diversified agriculture strategies and the sociocultural dimensions of marine aquaculture.

Index

Page numbers in italics refer to figures, maps, and tables

1978 Constitutional Amendment to Protect Important Agricultural Lands (IAL), 19, *20*, 23, 26

Abercrombie, Neal, 138–139
Agricultural Districts, 20–21, *22*, 26–31
Agricultural Lands of Importance to the State of Hawai'i (ALISH), 25–26
agro-tourism, 11, 24, 32, 163, 185–186, 189, 194, 213
Ahupua'a system, 3, 8, 45, 54–55, 59–63, 71–72, 78n. 1, 81. *See also* fishponds (loko i'a); Moku (land division); traditional Hawaiian crops
'Āina Momona, 54–55, 57, 64, 82
Alexander and Baldwin, 19, 29
amenity migration, 19, 28, 32
American Factors (Amfac), 19, 72
American Farm Bureau Federation, 172
Ancestral Visions of 'Āina (AVA Konohiki), 75–77
aquaculture, 40, 191. *See also* fishponds (loko i'a)
aquifers, 65, 68, 77
authenticity, 2, 5, 102–103, 186, 199
'auwai (water channels), 56, *58*, 60, 73

banana (mai'a), 3, 57, 60, 74, 167
beef, 3, 37, 159, 160, 178, 182
Big Five, 19, 29, 179; new, 165. *See also* Alexander and Baldwin; American Factors; Castle and Cooke
bioengineering. *See* biotechnology; genetically modified organisms (GMOs); genetic engineering (GE)
biofuel, 2, 32, 39, 142, 147
Biotech Industry Organization (BIO), 167
biotechnology: corporate, 122, 143–144; criticism of, 117, 123, 133n. 16; legislation of, 131n. 5, 151, 165–166, 213; place of, 117, 125, 129–131, 166–167. *See also* genetically modified organisms (GMOs); genetic engineering (GE); papaya; seed corporations
breadfruit ('ulu), 6, 44, 60, 74, 81
Bt (*Bacillus thuringiensis*), 143

Castle and Cooke, 19, 22, 29
Cayetano, Ben, 166–167
certification: GMO crop threat to organic, 123, 125, 133n. 15; food safety, 31, 94, 99; organic, 158, 160, 163, 189, 192; non-profit, 164, 192; seed, 145
chefs, 5, 103–104, 183, 193; farm collaboration, 11, 31, 188, 207. *See also* restaurants
civic agriculture, 86, 186
civic engagement, 1–2, 87, 97, 109, 111, 152, 211, 214

coffee: as commodity crop, 3, 204; GMO, 165; and labor, 193, 207–208; organic, 162; as specialty crop, 23–24, 26, 39, 43, 190
cold-chain logistics, 4, 182, 187, 209
College of Tropical Agriculture and Human Resources (CTAHR), 25, 38, 145, 149, 203
community: building, 6, 9, 172, 180, 197–198, 212; sense of, 3, 74, 93, 111, 186, 195
community education: and farmers' markets, 99, 108–111, 192, 209; and farms, 6, 209, 212; and fishponds, 9, 82–83
Community Food Security Plan, 46
community supported agriculture (CSA), 85, 159, 186, 188, 191, 211. *See also* MA'O Farms; organic farms
Conservation Districts, 21, *22*, 29
conventionalization hypothesis, 167, 171
corn, 141–142, 146. *See also* seed corn
Cultural Learning Center at Ka'ala, 6

dairy, 10, 37–38, 162, 170, 179, 193
Diamond v. Chakrabarty (1980), 143
Districts (Agricultural, Conservation, Rural, and Urban), 20–21, *22*, 26–31
diversified agriculture, 23–25, 138, 164, 167, 170; seed corn, 145–146; small-scale, 87, 94, 97, 111, 159, 212; strategies, 4, 14, 19, 23–25, 32, 210. *See also* specialty crops

Eat Local Challenge, 6
"Eat what there is" ('Ai i ka mea loa'a), 54, 76
Electronic Benefits Transfer (EBT), 48, 50, *92*, 96, 99, 107
energy, 13, 17, 20, 32, 49, 87, 142
E'o fishpond (Waipi'o, 'Ewa Moku, O'ahu), 68, *69*, 75

"fake farms," *20*, 27, 170. *See also* hobby farming
Farmers' Market Nutrition Program (FMNP), *92*, 106–107
farmers' markets, 5–6, 85, 174, 193–194; locations, 89, *91*; and low-income consumers, 9, 86, 105–107, 111, 113; and tourism, 9, 90, 104–105, 189, 213; types, 90, *91–92*, 94, 97, 100. *See also* farm security; food security
farmland preservation, 1, 12–13, 39, 43, 106, 150, 152, 206; policy, 8, 18–19, *20*, 21–22, 27–32, 138, 148
FarmLovers Farmers' Markets, 97, 100
farm security, 86, 102; vs. food security, 9, 39, 88, 105–107, 111. *See also* food security
fish, 13, 40, 74, 81–84, 94, 179. *See also* fishponds (loko i'a)
fishponds (loko i'a), 55, 57, *58*, 59, 64–68, *66*, *67*, *69*; destruction, 68, 72; restoration, 9, 68, 75, 77, 81–84, 212. *See also* Ahupua'a system; E'o fishpond (Waipi'o, 'Ewa Moku, O'ahu); Hanaloa fishpond (Waipi'o, 'Ewa Moku, O'ahu); He'eia fishpond (Kāne'ohe, Ko'olaupoko Moku, O'ahu); Ka'elepulu fishpond (Kailua, Ko'olaupoko Moku, O'ahu); Ka'ihikapu fishpond (Moanalua, Kona Moku, O'ahu); Kawainui fishpond (Kailua, Ko'olaupoko Moku, O'ahu); Kuapā fishpond (Maunalua 'ili, Waimānalo, Ko'olaupoko Moku, O'ahu); Kuhialoko fishpond (Wai'awa, 'Ewa Moku, O'ahu); Lelepaua fishpond (Waikīkī, Kona Moku, O'ahu); Moku (land division); Nu'upia fishpond (Kāne'ohe, Ko'olaupoko Moku, O'ahu); Pa'au'au fishpond (Wai'awa, 'Ewa Moku, O'ahu); traditional Hawaiian crops
foodbanks, 12, 47–48, 107
food democracy, 1–3, 5, 7–11, 13–14, 18, 211, 214
food deserts, 105–106
food insecurity, 1, 8, 11–12, 36, 45–48, 88
Food Safety Program, 31
food security, 1–2, 8, 12–13, 19–20, 31–32, 36, 49–50, 86; vs. farm security, 9, 39, 88, 105–107, 111; and self-sufficiency, 38–43, 212; and women, 156. *See also* farm security; food insecurity

Food Security Council, 50
Food Security Task Force, 50
food self-sufficiency, 5, 8, 13, 23, 40–42, 211–212; legislation, 37–38; traditional, 3, 36
fruit, 3, 74, 100, 179, 194; imported, 37–38, 101; locally grown, 4, 102, 170, 191, 207; organic, 162, 203; as specialty crop, 4, 23–25, 32. *See also* banana (mai‘a); breadfruit (‘ulu); noni; papaya; pineapple

gender: inequality, 13, 187; dynamics in organic agriculture, 156–159, 162, 169, 171, 173–174; roles, 182–183
genetically modified organisms (GMOs), 6, 39, 94, 157, 164–168, 213. *See also* biotechnology; genetic engineering (GE); papaya; seed corn
genetic engineering (GE), 131n. 1, 132n. 9, 133n. 15, 139–140, 143, 157. *See also* biotechnology; genetically modified organisms (GMOs); papaya; seed corn
GMO Free organizations, 124, 165. *See also* Hawai‘i SEED
Gonsalves, Dennis, 120–121, 199

Hanaloa fishpond (Waipi‘o, ‘Ewa Moku, O‘ahu), 68, *69*
Hawai‘i Agricultural Research Center (HARC), 134n. 19
Hawaii Beef Producers, 182
Hawai‘i Cattlemen’s Council, 126
Hawai‘i Community Foundation, 160
Hawai‘i Crop Improvement Association (HCIA), 138, 140, 145–146, 149–150
Hawai‘i Experiment Station, 118
Hawai‘i Farm Bureau Federation (HFBF): farmers’ markets, 90, *91*, 93–94, 96, 99, 102, 104, 106–107, 126; and organic agriculture, 160–161, 167
Hawai‘i Food Policy Council (HFPC), 50
Hawai‘i Papaya Growers Association, 126
Hawai‘i Papaya Industry Association (HPIA), 126, 134nn. 18–19
Hawai‘i Regional Cuisine (HRC), 5
Hawai‘i SEED, 124–125, 165
Hawai‘i seed corn industry (HSCI), 148–150
Hawai‘i State Civil Defense, 45
He‘eia fishpond (Kāne‘ohe, Ko‘olaupoko Moku, O‘ahu), 9, 75, 81–84
hobby farming, 27, 30, 168–170, 190
Hokuli‘a development, 27–28, 150
Ho‘oulu ‘Āina, 75
housing, 13, 23, 27–29, 32, 39, 150
hybrid crops, 121, 142–143, 145–146

Important Agricultural Lands Laws (IAL; Acts 183/2005 and 233/2008), 8, 19, *20*, 27–29
imported food: dependency on, 2, 4, 10, 17, 24, 40–43, 57, 111–112; disruption of supply, 43, 206; vs. locally grown, 5, 25, 32, 37–38, 112, 209; proportion of, 37, 132n. 12
import-substitution, 5, 25, 38
integration, vertical, 4, 24–25, 141, 144
intellectual property, 6, 130, 133n. 16, 143–144, 165

Ka‘elepulu fishpond (Kailua, Ko‘olaupoko Moku, O‘ahu), 65, *66*
Kahana Valley (Ko‘olauloa Moku, O‘ahu), 73–74
Kahekilinui‘ahumanu, 55
Ka‘ihikapu fishpond (Moanalua, Kona Moku, O‘ahu), 68
Kailua (Ko‘olaupoko Moku, O‘ahu). *See* farmers’ markets; fishponds (loko i‘a)
Kailua-Kona (Big Island), 204
Kako‘o ‘Oiwi, 75
Kamapua‘a, 61, 78n. 2
Kamehameha III, 70–71
Kamehameha Schools (KS), 75, 82–84
Kamilo Nui Valley, 39
Kanu Hawai‘i, 6
Kapapala Ranch, 182
Ka Papa Lo‘i o Kānewai, 6
kava (‘awa), 6, 77, 190
Kawainui fishpond (Kailua, Ko‘olaupoko Moku, O‘ahu), 65, *66*
Kawelo, Hi‘ilei, 9, 75, 82–84
Kealia Ranch, 182

Kōkua Hawai'i Foundation, 6
Kokua Kalihi Valley (KKV), 75
Konohiki, 55, 60–62, 64, 73
Kuahiwi Ranch, 10, 178
Kuapā fishpond (Maunalua 'ili, Waimānalo, Ko'olaupoko Moku, O'ahu), 55, 65, *67*
Kuhialoko fishpond (Wai'awa, 'Ewa Moku, O'ahu), 68, *69*

labeling, 105, 131, 133n. 15, 151, 182, 192
labor: cost, 4, 17, 39, 41, 187, 206, 209; human trafficking, 2, 185; immigrant, 40, 149, 179, 185, 187–188, 204; volunteered, 185, 196–200, 213–214; worker benefits, 188, 204, 206, 208–209. *See also* World Wide Opportunities on Organic Farms (WWOOF)
land: access to, 18, 31, 54, 146, 163–165, 206; concentration of ownership, 19, 71–72, 144, 150; prices, 20, 38–39, 119, 170, 172, 187; tenure, 13, 18, 45, 55, 60, 70–74. *See also* Agricultural Districts; Ahupua'a system; farmland preservation; housing; land use; Māhele (1848; division of lands); Moku (land division)
Land Evaluation and Site Assessment (LESA) Commission, 26
landscape: consumption, 8, 19, 32, 186; protection, 30, 32
Land Study Bureau (LSB), 25
land use, 13, 18, *20*, 21–22, 28–32, 140, 150–152, 206; transition, 2, 7, 55, 146, 148, 178; competition, 8, 32, 39. *See also* Agricultural Districts; Ahupua'a system; farmland preservation; fishponds (loko i'a); housing; land; Māhele (1848; division of lands); plantations; seed crops; urbanization
Land Use Commission, 22, 25–27, 29, 148
Land Use Law (1961), 19–22, 27–28, 30, 148
Lelepaua fishpond (Waikīkī, Kona Moku, O'ahu), 68
livestock, 10, 74, 81, 142, 159–160, 179, 180
local food: demand, 31, 106, 113, 183
localization, 1, 8–9, 41, 85–86, 88, 111, 156, 211–212
lo'i kalo (taro pondfields), 13, 55–57, *58*, 60, 62, 71–74, 77; restoration, 6, 75, 212. *See also* taro (kalo)

macadamia nuts, 4, 23, 25, 39, 43, 203–204, 207
Māhele (1848; division of lands), 3, 18, 54–55, 68, 71, 74
Makahiki, 60–62, 78n. 2
mala (dryland gardens), 59–60, 62
MA'O Farms, 6, 14n. 2, 75, 84, 101, 160
marker-assisted selection (MAS), 144, 147
McBryde Sugar Company v. Robinson (1973), 17, 72
medicinal plants, 75, 77
military, 68, 70, 75, 158; impact on food self-sufficiency, 40; martial law, 45
milk, 37, 179, 193. *See also* dairy
MJ Ranch, 182
Moku (land division), 59, 62; as aquifer management system, 65–66; of O'ahu, 62, *63*. *See also* Ahupua'a system; fishponds (loko i'a)
molecular biology, 10, 140, 143–144
monoculture, 138, 169; vs. polyculture (mosaic planting), 171
Monsanto, 122, 138, 143–144, 149, 165

Na'alehu Dairy, 179
Nalo Farms, 11, 160, 207
National Organic Certification, 158
National Organic Program, 158, 170–172
National School Lunch Program, 112
Native Hawaiians, 6–9, 13, 19, 21, 46, 119, 212; at farmers' markets, 90, *91*, 105–106; in anti-GMO movement, 6–7, 155–156. *See also* Ahupua'a system; fishponds (loko i'a); traditional Hawaiian crops
neoliberalism, 11–12, 212
niche market: locally grown food, 5, 32, 87, 146; organic food, 188, 191–192; tourism-oriented, 24, 186, 212
non-food agricultural products, 23–25, 32, 39–40, 43, 142
noni, 98–99

North Shore Cattle Company, 103
nutrition, 1, 36, 44, 86, 169; programs, 12, 48–49, 106–107, 112
Nu'upia fishpond (Kāne'ohe, Ko'olaupoko Moku, O'ahu), 65, *66*

Obama, Michelle, 14n. 2
Okimoto, Dean, 11, 160, 207–210
organic certification, 123, 125, 133nn. 15–16, 158, 163, 192
organic farming, 10–11, 124, 156–159, 161–163, 171, 213; conventionalization, 164, 171; women in, 156–160, 171–174
organic farms, 6, 11, 158, 160, 162; labor issues, 163, 185, 187–189, 193–194, 197–199, 214; large-scale, 167, 169–174, 188, 192, 203; volunteer experiences on, 194–197. *See also* Robb Farms; World Wide Opportunities on Organic Farms (WWOOF)
organic ideals, 157, 162–163, 173, 197–198
organic produce, 25, 103, 112, 159, 163, 188, 191
Otsuji Farms, 103

Pa'au'au fishpond (Wai'awa, 'Ewa Moku, O'ahu), 68, *69*
Paepae o He'eia, 9, 75, 81–84
Pahoa, 119
Palikū, 61
Papahana o Kuaola Hui Kū Maoli Ola, 75
papaya, 103, 118, 204, 208; organic, 117, 123–127, 129, 133nn. 15–16, 213; transgenic, 6, 9, 116–117, 121–131, 133nn. 15–17, 134n. 19, 165–167, 213; varieties, 118, 122, 132n. 7, 132n. 10
Papaya Administrative Committee (PAC), 121–122, 128, 132n. 10, 134nn. 18–19
Papaya Ringspot Virus, 117, 119–125, 127–128
parasite-derived resistance, 121, 132n. 9
People's Open Markets (POMs), 90, *91–92*, 94–97. *See also* farmers' markets
pesticides, 79, 139, 150, 157–158, 213; drift, 13; residue, 68, 158
pests, 117, 120, 132n. 14, 141, 143, 148, 163
pig (pua'a), 74, 81, 125; as body-form of Lono, 61. *See also* Kamapua'a
pineapple, 4, 25, 194; marketing, 23. *See also* plantation agriculture; plantations
place-identified foods, 5, 23–24, 132n. 7
plantation agriculture, 3–4, 9, 17, 19, 37; decline of importance, 19, 26, 119, 123, 138, 146, 150, 164; labor, 145–146, 185, 188
plantations, 3–4, 17, 26, 71–72, 148, 179; closure, 4, 22–23, 100, 119, 123, 164, 170, 179
plant breeding, 139–145
pocket market problem, 4, 24, 187
pollination, 44, 142, 145, 148–149; GMOs and cross-pollination, 126, 130, 133n. 15, 164
privatization: and Māhele, 9, 18, 55, 71, 74; and neoliberalism, 11

restaurants, 163, 172, 178, 197, 207, 210, 213; food safety, 39; upscale, 5, 6, 11, 198. *See also* chefs
rice, 3–4, 14n. 1, 37, 45
RJ Ranch, 182
Robb Farms, 11, 203–206
Rodale, J. I., 158, 169, 203
Rural Districts, 21, *22*, 29

second homes, 19–20, 28, 30–31. *See also* hobby farming; land use; tourism
seed corn, 7, 9–10, 121, 138–152, 193, 208, 213. *See also* seed crops
seed corporations, 7, 9–10, 122, 138–140, 143–144, 147–152, 161, 165; (in)visibility of, 150. *See also* Monsanto; Syngenta
seed crops, 32, 39, 164–166, 168, 213. *See also* seed corn
self-sufficiency, 2, 13; and Ahupua'a system, 3, 76; and food democracy, 8–9, 36, 40–42, 212; low level, 5, 123; as policy goal, 23, 37–38, 211. *See also* farm security; food security
self-sufficiency bill (HB2703 HD2, 2012), 37–38, 41

senior citizens (kūpuna), 47, 64, 209; at farmers' markets, *91*, 96, 100; food preferences, 76, 83
shipping, 4, 76, 85, 182, 187; costs, 24, 38; danger of disruption, 8, 17, 43–46, 206; energy use, 87
social media, 94, 97, 104
soil, 163, 178, 198; conservation, 124, 157; improvement, 163, 198; in land assessment, 25–26; microbes, 203
specialty crops, 4, 23, 25, 188–190, 204; herbs, 37, 39, 43; salad greens, 5, 103, 197, 204, 207; vegetables, 4, 74, 100, 103, 204. *See also* coffee; diversified agriculture; fruit; macadamia nuts; niche market; non-food agricultural products; organic farming
standards, 23, 26, 29, 102, 192; testing, 125, 133n. 15, 165. *See also* certification
statehood (1959), 19–20, 68, 72
State of Hawai'i Departments: of Agriculture (DOA), 25–26, 119–120, 125, 129, 132n. 12, 133n. 15, 161, 167; of Business, Economic Development and Tourism (DBEDT), 50, 104; of Education (HI DOE), 112; of Health (HDOH), 39, 46, 82; of Transportation (DOT), 109–110
subsidies, 30, 88, 97, 141, 170
subsistence farming, 3, 19
sugar (kō), 3, 17, 56–57. *See also* plantation agriculture; plantations
Supplemental Nutrition Assistance Program (SNAP), 48–49
sweet potato ('uala), 44, 56, 60, 74, 81
Syngenta, 138, 144, 165, 167

taro (kalo), 6, 26, 37, 59, 73, 78; body form of Kāne, 59; and genetic modification (GM), 165–166; and Kamapua'a, 61, 78n. 2; leaves, 74. *See also* lo'i kalo (taro pondfields)
tax: and indigenous landholders, 72; provisions for agricultural land, 21–23, 28, 30, 170; revenue, 150; subsidies, 88
tomato, 5, 74, 103, 111, 159, 210
tourism, 5, 9, 11, 19, 28, 72, 145, 214; dependency on, 17, 180, 213; and farmers' markets, 86, 90, *91–92*, *93*, 98, 101–105, 111. *See also* agro-tourism; amenity migration; World Wide Opportunities on Organic Farms (WWOOF)
traditional Hawaiian crops, 44, 55–57, 76, 81; as body forms of akua, 61. *See also* banana (mai'a); breadfruit ('ulu); noni; pig (pua'a); taro (kalo); sugar (kō); sweet potato ('uala)
transgenic crops. *See* corn; papaya; seed corn
transportation. *See* shipping

University of Hawai'i, 75, 120–121, 129, 145; lo'i kalo, 6. *See also* College of Tropical Agriculture and Human Resources (CTAHR); Land Study Bureau (LSB)
Urban Districts, 21, *22*, 27, 29
urbanization, 22, *22*, 74, 119; and aquifers, 68; counter-urbanization, 28; expansion, 8, 19, *20*, 27; sprawl, 13, 29–30. *See also* housing
US government agencies: Animal and Plant Health Inspection Service (APHIS), 121; Department of Agriculture (USDA), 46–48, 86, 112, 133n. 15, 157–158, 167, 170, 192; Environmental Protection Agency (EPA), 121; Food and Drug Administration (FDA), 121, 167; National Institutes of Health (NIH), 143; National Science Foundation (NSF), 143

vegetables, 4, 32, 38, 74–75, 107, 162, 170; imported, 32, 37, 101; locally grown, 2, 24, 95, 159, 191; organic, 188, 203; temperature sensitive, 204. *See also* farmers' markets; specialty crops; traditional Hawaiian crops
volunteered labor, 73, 82–84, 123. *See also* World Wide Opportunities on Organic Farms (WWOOF)

wages, 12, 160, 188, 204; living wage, 189, 198, 205, 209
Waiāhole, 71–72
Waianu Farm (Big Island), 160
Waikīkī, 45, 56–57, *58*, 72, 145; Ahupuaʻa, 68
Waimea (Big Island), 194, 203
Waimea Valley (Oʻahu), *98*, 110
warehouse stores, 85, 88, 95, 170, 197
water, 13, 31, 54, 165, 178; diversion, 13, 17, 71–72, 146, 148; as public trust, 72; quality, 82, 158; supply, 38, 44, 56; traditional management, 54–55, 59–62. *See also* Ahupuaʻa system; aquifers; fish; fishponds (loko iʻa); loʻi kalo (taro pondfields)
wetland agriculture. *See* loʻi kalo (taro pondfields); rice
Whole Foods Market, 85, 109, 160, 170, 204
WIC (Special Supplemental Nutrition Program for Women, Infants, and Children), 48–49, 107
women, 10, 64, 156–160, 162–169, 212; marginalized role in agriculture, 159
Worldwatch Institute, 42
World Wide Opportunities on Organic Farms (WWOOF), 10–11, 163, 185–187, 189–191, 193–200; as start-up strategy, 198, 213–214

zoning, 20–22, 27, 170, 206; redistricting petitions, 22; rezoning, 106, 112. *See also* land use